SCIENCE
D0215812

ENOUGH OF PESSIMISM

Philip H. Abelson

ENOUGH OF PESSIMISM

100 Essays

by

Philip H. Abelson

AMERICAN ASSOCIATION FOR
THE ADVANCEMENT OF SCIENCE
Washington, D.C.

AAAS publication 85-5

Frontispiece by Yoichi R. Okamoto

Library of Congress Cataloging in Publication Data
Abelson, Philip Hauge.
Enough of pessimism.
1. Science. 2. Engineering. 3. Technology.
I. Title.
Q158.5.A34 1985 500 85-3924
ISBN 0-87168-310-5
ISBN 0-87168-274-5 (pbk)

Distributed internationally by
Birkhäuser—Boston, Basel, and Stuttgart

Printed in the United States of America
First Edition

Contents

Foreword

Recruiting Philip Abelson as editor of *Science* was one of the most scientifically fruitful and personally satisfying accomplishments of my years as executive officer of the American Association for the Advancement of Science. His predecessor, Graham DuShane, had fended off several invitations to return to an academic chair, but in 1962 he received an offer too challenging to resist; he told us he planned to join the Vanderbilt University faculty as dean of graduate sciences. When the AAAS board of directors asked me to recommend a successor I did: Philip Abelson.

The *Journal of Geophysical Research* was flourishing under his leadership. He and I had had several discussions about editorial problems. I knew he was interested not only in the intellectual content of *JGR* but also in the mechanics of editing, in speeding up the peer review process, and in searching for appropriate articles. Moreover, and of great importance in selecting an editor for a magazine with as diverse scientific content as *Science*, his own interests were very broad. In a biographical account that was later published in *Science* to introduce the new editor, Frank Campbell reported that when Abelson had been elected to the National Academy of Sciences, he chose to be enrolled in the geology division but that he might have selected with equal appropriateness any of five other divisions.

After two or three meetings to discuss *Science*, AAAS, and the opportunities they presented, Abelson said yes, he would accept the editorship, but with one important reservation: He did not wish to write editorials.

During Graham DuShane's regime he and Joseph Turner, the associate editor who had recently resigned, had been writing two-thirds of the editorials. Guest authors and I provided the rest. Reluctantly I decided I could write more and we could find more guest editorials, so we agreed, and since 3 August 1962, the magazine has borne the name of Philip Abelson as editor.

In those first four months, he used the editorial page only once; as other editors had before him, he described his plans for *Science.* In most other weeks, however, editorials came with tiresome frequency over my initials. Perhaps that frequency helped motivate him to tell me one welcome day that he had decided he would like to try his hand at writing editorials.

It was a skillful hand, as readers quickly learned from his first three editorials, all of which appeared in December of 1962: "The Venus Mission" (page 16), "Civilian Nuclear Power" (page 110), and "Science and the Humanities" (page 196). Those three gave a solid foretaste of the many others that came from his pen over the next twenty-two years. They ranged over a variety of topics; there was never any telling from one what the subject of the next might be. They illustrated one of his editorial tactics, that of using a recent major report as an opportunity for analysis and comment on the primary issues involved. But Abelson went beyond the typical newspaper editorial style of expressing and justifying an opinion to provide some solid information. Readers not only learned his opinion but also gained some practical knowledge or better understanding of the nature of the issues involved. For example, in his first editorial, on the Venus mission, he explained how often and by how much the Venus probe would change speed on its 182-million-mile flight from launch to its nearest approach to Venus.

In the last issue of *Science* that bore his name as editor, Graham DuShane wished his successor a long and distinguished career. That wish has been fulfilled, and amply so, not only through his other responsibilities but also as a distinguished and prolific editorialist whose understanding of science, technology, policy, and their interrelationships is well illustrated by the 100 samples selected for this volume from among more than 450 he wrote for *Science* during his tenure as editor. Happily, the list is not complete; although leaving the editor's chair, he has agreed to continue to add to the list of Philip Abelson editorials.

Dael Wolfle
26 November 1984

Introduction

It has been my good fortune to have worked for and with Phil Abelson for nearly a decade and a half. I recall clearly my initial impression of the man: I had flown to Washington to talk to him about spending a postdoctoral year writing for *Science*. He enthusiastically pressed into my hands the 400-page lunar science issue, hot off the press, and gruffly inquired what I had done my thesis research on. We talked about science, not about writing (for which I was grateful, since I had no experience in writing or journalism to speak of). And when the conversation came finally to my wish to spend a year at *Science*, he was silent for what seemed an eternity—a tactic that I later came to know as vintage Abelson. When he finally said yes, I had the distinct impression that it was in large part because he was too kind to say no.

Those early impressions have endured. As an editor, he provided advice with economy; he didn't waste words on a manuscript but went right to the central points. He often chose to communicate this advice by letter or phone, for he found it easier to be direct at a distance than in person. The fierce, iconoclastic editor—Abelson the Terrible I had heard him called—turned out to be a myth behind which a reserved person could take refuge. And his enthusiasm—for jogging, for good wines, for bridge, for his tomato garden, and especially for science—is more than an occasional mood; it is his hallmark.

Just as his enthusiasm for science was wide ranging, he responded to the enthusiasm of others. He rarely delved into a field, no matter how remote from his own background, without encountering new results or techniques that he found fascinating. He reserved particular enthusiasm for experimental results derived from pushing a technique to new limits and was quick to see the implications for further research or application. He was eager to share the results and their significance with others.

From enthusiasm came a basic optimism. His years in the editor's chair and in Washington bred a certain cynicism about the promises of authors and of governments, but it did not seem to corrupt his faith that problems were soluble if reason was applied. And he had, characteristically, a pragmatic rationale for this philosophy: In "Enough of Pessimism" (page 12) he holds out the examples of Franklin and Jefferson in both science and politics, noting that "It is the optimists who achieve."

As I came to know Abelson, however, it became apparent that his success as an editor owed a lot to several other characteristics as well—his quickness and toughness of mind, his willingness to set aside preconceptions and think things through from first principles. To work beside Abelson was to watch how rapidly he grasped the essentials of a new scientific idea, turned it this way and that to see its implications, and probed it for its weaknesses. When he found them or detected salesmanship, he was quick to doubt and willing to do so publicly, even when it was profoundly unpopular with his constituency. When his enthusiasm led him to endorse an idea or a field that later developed problems, he was ready to accept the new evidence and form a new opinion. To give a personal example of his open-mindedness, I very much doubt that *Science 85* would exist if he had been unwilling, at a critical stage in its development, to look carefully at a new and unproven idea and support it on its merits.

Abelson has been an acute observer and recorder of what was going on around him during his editorial tenure that spanned 22 years of *Science*'s 103-year history. His editorials map the progress of science, the evolution of our society, and the changing interaction between them. You can follow in them the swings and counterswings of opinion on several of the great technological efforts of our times—nuclear power, the space program, industrial competition with other nations. Often, his editorials go further than the popular opinion of the day to anticipate the core of the issue—that which would remain a problem for a long time. It is the record of an inquiring mind, a distinguished intellect, and a great editor.

Allen L. Hammond
5 January 1985

ENOUGH OF PESSIMISM

The Bicentennial year is an appropriate time for comparisons between earlier times and now. In terms of knowledge, education, affluence, and health the contrast is great: there has been substantial progress. However, in terms of leadership and morale, the opposite is true. The people of those times were rich but we are poor. They had leaders of stature, breadth, and vision, who in keeping with the spirit of the times, faced the future with faith and optimism. In comparison to a Franklin or a Jefferson, our leaders and would-be leaders seem only ordinary. Caught up in the excitement of the present, they rarely look beyond the next election.

Who among the present-day politicians can come close to matching Franklin's enthusiasm, foresight, and knowledge of the world around him? These qualities were exemplified when he wrote in 1780:

> It is impossible to imagine the height to which may be carried, in a thousand years, the power of man over matter. We may perhaps learn to deprive large masses of their gravity, and give them absolute levity, for the sake of easy transport. Agriculture may diminish its labour and double its produce: all diseases may by sure means be prevented or cured . . .

And who among our politicians comes close to the breadth of a Jefferson who, though a successful lawyer, was an avid student of nature, a talented botanist and paleontologist with a deep interest in all the other sciences? What politician today would have either the imagination or the convinced insight to make a statement matching Jefferson's, "knowledge is power, knowledge is safety, knowledge is happiness"?

Our poverty goes beyond a lack of leadership. It extends to a malaise of the spirit of our people. Indeed, such is the pathology that, even if the Messiah should appear, he or she would either go unrecognized or, if recognized, would soon be chopped down to size. At the same time this country has turned its back on optimism and is becoming a nation of pessimists.

During most of the country's history, perhaps its greatest assets were its faith in progress, its can-do spirit. Sometimes exu-

Enough of Pessimism

berance was overdone but better that than the opposite, as any experienced scientist can testify. The research worker who is convinced ahead of time that experiments will be fruitless seldom is proved wrong in that judgment. It is the optimists who achieve.

In this country optimism was at its peak early during times of great poverty, hardship, and amid unmerciful ravages of disease. But Franklin's optimism was justified by events. Great increases in knowledge and enormous improvements in agriculture, medicine, and technology liberated many humans from much of the drudgery and pain that had previously been their lot. But the behavior of humans is weird and wonderful. Far from feeling gratitude toward benefactors or admiring the great edifice of knowledge that makes their comforts possible, they have now turned sour and their attitudes are reflected by their chosen representatives.

Part of their feeling toward science may be due to another factor. Shortly after World War II, public opinion accorded science a high place in the scheme of things. For nearly twenty years science was exalted in the press, by the public, and by politicians. Expectations were aroused that could not be fulfilled. A swing of the pendulum was inevitable and it has been going on for about ten years. The public has the impression that scientists, engineers, and physicians are not delivering the perfect performance that should be expected of them. At the same time relatively small side effects of new technology and medicine have appeared. In view of the insatiable need of the mass media for stories, the seriousness of these effects has been greatly exaggerated. The backward swing of the pendulum has also been abetted by some scientists who have been leaders in creating more problems and more pessimism than the facts justified.

Pessimism is a kind of sickness that debilitates the individual and the country. One would not advocate that we become a nation of Panglosses. However, enough of pessimism. It leads nowhere but to paralysis and decay.

FRONTIERS

From his first editorial, Philip Abelson has been interested in the frontiers of research—in space, beneath the sea, and in the lab. His first question to a scientific colleague encountered at a conference or over the telephone is invariably, "What's new?"

He pointed out early the importance to astronomy of new regions of the electromagnetic spectrum, such as radio and infrared. He has repeatedly stressed the importance of new scientific instruments for progress in research and observed the coming of age of the computer, one of the more important contemporary instruments.

Abelson's deep interest in the earth sciences reflects to some degree his own research in geochemistry, but extends far more widely than that. In the midst of the scandal over the collapse of early efforts to drill into the seafloor, he proclaimed the enormous potential of drilling—a potential later fulfilled by its contributions to the plate tectonic synthesis. He championed the new fields of earthquake and volcano prediction, while cautioning that true predictive ability would require decades of work. He pointed out that the impact of asteroidal bodies on Earth opens up a major new field of investigation, including their still controversial role in large-scale extinctions.

He has been keenly aware of and supportive of planetary and lunar science. He celebrated the analysis of the lunar samples and the remote exploration of Venus, Mars, Saturn, and other planets as major advances in our knowledge. But he has not hesitated to question skeptically the search for life on Mars or grandiose plans for manned missions when he thought unmanned satellites could suffice at a fraction of the cost.

He has also been interested in the physics and chemistry of novel materials, from polymers to crystals to composite substances, pointing out their remarkable technological and economic potential. And he has pointed to the research opportunities in fields ranging from plant science and cancer research to catalysis and chemical reactivity, noting the new frontiers opened up by instruments such as lasers, molecular beams, and ion cyclotron resonance.

In field after field across an extraordinarily broad spectrum he has shown a deep regard for the science, for what could be learned.

7 December 1962

The Mariner II mission is already a record-breaking success. The precalculated flight trajectory has been followed, all interplanetary experiments have functioned, and many engineering data have been acquired. Though *Mariner II* is now more than 23 million miles away, data from 90,000 measurements a day are being received.

The prerequisite for a successful spaceflight is functioning of all components. During launch phases, vibration and acceleration place unusual stresses on the vehicle. Even partial failure of one of hundreds of thousands of components can nullify the performance of all. After ascent to a circular parking orbit 115 miles from the earth, *Mariner II* was allowed to coast to a calculated point and was then boosted to escape velocity. During the next eight days the spacecraft was tracked to determine its path, and a slight corrective maneuver was made. The magnitude of the guidance problem can easily be seen. When *Mariner II* misses Venus by 21,000 miles next week, it will be 26.3 million miles from Earth. The spacecraft will have traveled 182 million miles at highly variable speeds. Starting from rest with respect to Earth, the velocity rose quickly to 18,000 miles per hour. The vehicle was later accelerated to 25,503 MPH, a speed in excess of escape velocity. After three days the velocity had decreased to 6874 MPH. Then the spacecraft was moving about the sun 6874 MPH slower than the earth's 66,000 MPH, that is, about 59,400 MPH. From that time the velocity of *Mariner II* increased as it moved toward the sun. The craft will attain a velocity of 84,000 MPH and catch up with Venus, which moves about the sun at 78,300 MPH. These figures make evident the complexity of calculating the trajectory and attaining it; this is only one facet of a successful flight.

The experimental and engineering data sensors must operate, and their information must be transmitted back to Earth. Hence the spacecraft must be positioned so that the solar batter-

The Venus Mission

ies can operate and the antenna is directed toward Earth. Miraculously, all the components of *Mariner II* have functioned.

Prospects are excellent that worthwhile measurements will be made during planetary approach. At that time two additional instruments will be turned on, a microwave radiometer and an infrared radiometer. They should measure temperature distribution on Venus and tell whether there are discrete clouds with breaks. Previous measurements from Earth seem to indicate a surface temperature of 300°C, but this value is not universally accepted.

The magnetic field of the planet will also be measured. If it is comparable with that of Earth, the observation will be interpreted as indicating that Venus has a hot molten core. Other similarities of Venus and Earth such as a history of differentiation would also be inferred.

The striking success of the Mariner II mission is reassuring. We now have grounds to hope that the Space Administration will ultimately shake down into an organization capable of sponsoring and carrying out solid scientific research.

---

4 December 1964

Astronomy is in the midst of a vital era according to an excellent report entitled "Ground-Based Astronomy: A Ten-Year Program" that has been prepared by the National Academy of Sciences. During the 1950s came the discovery by optical methods that the abundance of heavy chemical elements varied from star to star and was related to stellar age. This finding suggested that the elements were continuously synthesized in the stars and distributed by stellar explosions. Other major discoveries have come from radio astronomy. One of these is the observation that some galaxies emit great quantities of radio energy. The radio emission is due to synchrotron radiation resulting when high-energy electrons encounter a magnetic field. Associated with the electrons must be heavy particles. Thus, the discovery of these radio explosions shows that magnetic fields exist in interstellar space and that large numbers of high-energy particles are moving through these fields.

Simultaneous optical and radio observations recently have led to identification of a new class of astronomical objects. These are quasi-stellar sources whose presence was originally signaled by strong radio emission. In optical studies the objects appear similar to ordinary stars. The radiations have large red shifts indicating that distant objects are radiating energy at an enormous rate. Calculations suggest that a new type of energy source is required—perhaps the release of energy stored in a gravitational field of a collapsing body.

As a kind of sideline to consideration of these great cosmological phenomena, ground-based astronomy has made significant discoveries concerning our solar system. Recent optical studies have yielded new information about the atmosphere of Mars. A total atmospheric pressure of about 25 millibars is indicated, with about 14 microns of precipitable water. Warm spots on the moon and unevenly hot regions in the atmosphere of Venus also have been observed. Radio astronomy has yielded a more accurate measurement of the distance to the sun, estimates of the composition and roughness of the lunar surface, and of the temperature and structure of the surface of Venus. Fluorescent radiation from the lunar surface has been observed. Spectroscopic study of this light promises to reveal the nature of attendant chemical events. These contributions relate importantly to the space effort. For instance, the new estimates of the density of the Martian atmosphere have affected designs of equipment for landing capsules.

The potentialities of existing equipment have been only partially exploited. Improvements in auxiliary equipment are opening new opportunities. It was exploitation of new infrared detectors that led to the discovery of warm spots on the moon. Development of new devices such as image tubes promises extension of the effective range of telescopes.

A conspicuous potentiality in radio astronomy is the improvement in resolution to be obtained by constructing extended arrays of moderate-size radio telescopes. Such an array could detect and resolve radio sources even at the bounds of the observable universe.

The total cost of a ten-year program designed to provide important new facilities and additional astronomers has been estimated at $227 million. The Academy is conservative when it states, "... an investment in ground-based astronomical facilities of the order of one-half of 1 per cent of that going into the space effort would be consistent with a balanced program of federal support for science."

20 August 1982

A recently issued report [*Astronomy and Astrophysics for the 1980s* (National Academy Press, 1982)] begins: "Nature offers no greater splendor than the starry sky on a clear, dark night. Silent, timeless, jeweled with the constellations of ancient myth and legend, the night sky has inspired wonder throughout the ages."

For most of human history, leadership in studying the heavens has resided elsewhere, but during the twentieth century the United States has been the world center of astronomy. This preeminence has been due to good financial support and the imaginative creation of innovative observing equipment. The capabilities of excellent optical telescopes, developed during the first half of this century, were later extended by equipment designed for observing throughout the electromagnetic spectrum. Leading supporters of the development of optical telescopes were the Carnegie Institution of Washington, with its 2.5-meter telescope at Mount Wilson, and the Rockefeller Foundation, which gave the California Institute of Technology funds to build the 5-meter telescope at Mount Palomar. More recently, the National Science Foundation has become a major funder of ground-based astronomy, while NASA has provided excellent facilities in space. The United States has led in exploration of the solar system. In addition, it has launched space vehicles that have permitted observations that could not be achieved from the earth because of absorption of radiation in the atmosphere. The Space Telescope, to be launched in several years, will be free from atmospheric inhomogeneities that blur sources of light and will be capable of high resolution of objects.

By 1970, generous support of American astronomy had led to many discoveries, including Hubble's expanding universe, time and celestial distance scales, quasars, x-ray sources, high-energy celestial gamma rays, the cosmic microwave background radiation, and polyatomic molecules in interstellar clouds. Discoveries during the 1970s included neutron stars accreting matter from nearby companion stars, hot intergalactic gas whose mass rivals that of the galaxies themselves, vast regions of inter-

Astronomy and Astrophysics in the United States

stellar gas heated to hundreds of thousands of degrees by shock waves from supernova explosions, and a gravitational lens effect observed as the splitting of light from a distant quasar as the light passed through an intervening galaxy.

The contributions of American astronomy are important and impressive. However, leadership cannot be maintained by resting on our laurels. Continuing preeminence of the United States will be dependent on well-trained people who are provided with superior equipment.

The astronomical community has made a careful and searching study of opportunities and needs for support for the 1980s. Through extensive consultation and deliberation, a consensus has been achieved. The major new equipment recommended includes an Advanced X-ray Astrophysics Facility operated in space; a Very Large Array of radio telescopes; a new technology telescope, 15 meters in diameter, for ground-based studies in the optical and infrared regions of the spectrum; and a large deployable reflector in space. All of these proposals would substantially extend the capabilities of astronomy. For example, the Very Large Array would have an angular resolution one hundred times better than that of any other image-forming telescope at any wavelength. It would yield detailed radio images of quasars, the nuclei of galaxies, and features of interstellar molecular clouds and other astronomical objects. The first and third items above would be important for many studies, perhaps the most interesting being the examination of extremely distant objects whose radiation was emitted early in the history of the universe.

The report is well constructed and readable. It states well the case for additional expenditures for astronomy. Because of current budgetary problems, its recommendations may not be quickly accepted. However, it is designed to be relevant to the 1980s and at least part of it will surely be ultimately implemented.

15 July 1966

The relation of humans and computers has entered a new era, in which interaction is becoming quick and simple. A decade ago, use of computers seemed impractical for most scientists. Those conversant with the machines talked of the frustrating hours spent in programming, compiling, and debugging. There were long delays between concept and fruition, as users awaited their turn. Usually the computations merely took advantage of the fact that the new devices were much faster than desk calculators.

In this last decade the speed and memory of computers has been improved. Present-day models calculate 10^7 to 10^8 times as fast as a human does. These developments, however, are not so important as those that make for greater ease of using the machines. A sizable group of very competent men and women have created libraries of programs and subroutines. Some of them have turned their attention to other quick means of communicating with computers and to the complementary problem of obtaining information from the machine in a much more usable form.

An impressive sample of this progress was presented at a recent symposium at the Bell Telephone Laboratories in Murray Hill, New Jersey. The audience numbered about two hundred and included representatives from more than a hundred universities throughout the country. The group was truly interdisciplinary; it included substantial representation from the humanities, the social sciences, biology and medicine, the physical sciences, and, of course, mathematics and engineering.

At the Bell Laboratories some fourteen hundred individuals spend at least half their time working with computers, of which forty are available. These include special-purpose computers for on-line problem solving, a console for handling pictorial input and output, and a computer that takes graphic or symbolic input and delivers auditory output. Under design is a system in which a central machine will serve two- to three-hundred typewriter consoles and contain an elaborate program library.

Today, instructions to a computer can often be conveyed by typing simple English or abbreviations. Another means of easy access to the computer is through a console that employs something similar to an oscilloscope screen and a light pen. The computer converts a rough sketch into a finished drawing. Block circuit drawings can be quickly assembled through a series of instructions mediated by a light pen.

The computer can communicate with the user in new, simple forms. To the scientist, perhaps the most impressive development is the graphic presentation of data. Earlier, the output from the machine usually took the form of almost indigestible quantities of printed results. Today a glance at a curve on a screen or the plot of a thousand points can provide an almost instant summation of the same output. Another impressive development is that of teaching a computer to talk. Through manipulation of controls, the investigator can change the character and emphasis of the speech.

This work seems to be speeding the day when it will become possible to speak to a computer and to obtain quickly a spoken as well as a visible output. Improvements in computers and in the ease of using them portend a further great expansion in their use in all the sciences and in many of the humanities. One participant at the symposium remarked, "After growing wildly for years the field of computing now appears to be approaching its infancy."

10 December 1971

To a large extent, American leadership in science has been based on the widespread availability of excellent instrumentation. In an earlier era, scientists could make fundamental discoveries with the equivalent of sealing wax and string. Today an occasional worthwhile observation is made with simple tools, but most significant advances depend on the application of complex instrumentation. In many instances appropriate devices make possible a tenfold or greater speed in data collection. In other instances sophisticated equipment permits measurements and experiments heretofore inaccessible. Current trends indicate that, in the future, leadership in science will be even more contingent on pioneering the use of new and increasingly powerful equipment. American scientists are fortunate in having the support of an innovative instrumentation industry that has been a by-product of federal support of research.

The grants system placed considerable sums of money at the disposal of a large number of investigators who were a good market for effective apparatus. Many small companies were organized to invent, develop, and manufacture new products. Some companies produced unneeded or shoddy goods, and they failed. Other built needed and excellent equipment that was crucial to the advancement of science. Academic and industrial research benefited alike, and a thriving export trade was established.

One can learn something about economic systems and about the role of instrumentation in science by considering the contrasting situation in Russia. Those who have visited Russian laboratories generally come away with a favorable impression of individual scientists, their interest in science, their willingness to work, their familiarity with the literature, and their eagerness to learn. Yet much of the Russian work seems pedestrian. In many areas the Russians are followers, not leaders, despite the fact that large staffs are active. The consensus of visitors is that a major Russian deficiency is in their equipment. The creative potential of many fine young people is lost, for they must devote their time to making routine observations that could be made much faster and more accurately with modern equipment. At some institutes one may observe individual pieces of apparatus that have been invented and built there. However, scientists at other institutes seem unable to benefit from the inventiveness of their countrymen, for in the U.S.S.R. there is no adequate scientific instrumentation industry. In Western Europe the situation is quite different. Many laboratories are well equipped, and they have available the backup of inventive companies.

In view of the key role of instrumentation in the progress of science, policies with respect to allocation of federal funds have been shortsighted. Several years ago when the budgetary squeeze became severe, the National Science Foundation implemented a policy of favoring support for personnel over support for instrumentation. At the universities individual grantees followed the same practice. The fraction of funds devoted to instrumentation was not very large, so that their diversion did not help the employment situation very much. However, in the absence of adequate replacements and the purchase of new kinds of instruments, the quality of equipment at many universities has declined, and research is being hampered. The deficit in scientific equipment should be met, and federal authorities should establish a long-term policy of steady support for the procurement of instrumentation on a level that will guarantee continued American leadership in science.

7 February 1964

The lengthy controversy concerning Mohole drilling had some constructive aspects. Strong-minded, knowledgeable scientists cared enough and believed enough in deep-sea drilling to fight over how it should be conducted. Disagreement over the proper scope of the project led to recognition of the desirability of a broad program of exploration of the rocks under the oceans. Even the spectacle of scientists in conflict had a positive side, for the resultant open discussion provided the Congress and the public with a basis of trust in the eventual decisions. But a good thing can be overdone. There has been more than enough controversy about Mohole. We should all be pleased with the way Leland Haworth, director of the National Science Foundation, has assumed authority in the matter, and with the prospects for an effective future organization.

The time has come to forget the trauma of the past and to look toward the opportunities of the future. We know so little about the rocks under the sea that the potentialities are difficult to evaluate. But that is precisely the reason we should explore them. In the scientific area, many major possibilities have been pointed out that are of interest to everyone, for drilling at sea may provide information on the history of the earth, the origin of life, the evolution of life, continental drift, and the emergence and submergence of continents. One of the possible results of such investigation would be of both scientific and practical importance. This is the elucidation of the geochemical processes that have occurred and of the resultant concentrations of chemicals to be found in the rocks. On continental shelves, exploration and exploitation are already proceeding.

Some of this activity can be carried out by techniques already developed by the petroleum industry. Most of the sea bottom, however, is not accessible to such techniques. In attempting to assess what might be found under the deep seas, there are two possible approaches. One is to look at the material obtained from the ocean floor. These samples are usually of only moderate interest, though there are exceptions, such as the manganese nodules. A second approach is a more general one. Most of the rocks at the surfaces of the present continents were formed from sediments deposited in marine environments. We can use them as a kind of rough yardstick to gauge what might be found under the present seas. In the past, the oceans have played the role of a gigantic laboratory in which chemical processes have led to concentration of relatively rare substances, both organic and inorganic, into commercially valuable occurrences. An example is the role of the marine environment in the formation of petroleum. Concentrations of trace elements in the black shales ultimately will have great economic significance. On the continents, exploitation of the crust is destined to provide returns of the order of trillions (10^{12}) of dollars.

Since the seas cover about 71 percent of the earth's surface, the wealth under them could be greater than that now known on land. To obtain even a glimpse of the scientific and economic potential will involve great effort and the application of our best scientific and engineering talents. We should not make small plans or squabble further about objectives. Both deep and shallow exploratory drilling are worthwhile. Both should be carried forward with no more delay.

25 May 1973

We live on a restless planet. Continents are in motion, as they have been for at least 130 million years and as they are likely to be for a long, long time. These motions give rise to earthquakes, notably around the rim of the Pacific Ocean, but also in other areas. During this century, earthquakes have killed hundreds of thousands of people and caused tremendous property damage. Most are of small intensity, and these occur frequently. In a given locality, the really large events come at intervals of perhaps fifty years or more. Experts believe that, by the end of this century, California will probably experience a killer earthquake causing as much as $20 billion in damage.

After large earthquakes in the past, there have been in-depth studies of the event and its sequel (for example, the Alaskan earthquake in 1964). However, studies aimed at prediction have enjoyed relatively less attention. We have learned some lessons from past observations, but we must learn much more if we are to minimize future damage and loss of life.

One desirable goal is ability to predict both the timing and the intensity of major earthquakes. Recent research has been moving us closer to this goal. Earlier work had indicated that premonitory events precede earthquakes. There have been reports of a change in frequency of occurrence of small local earthquakes preceding a large one. Other effects noted have included changes in tilt, fluid pressures, radon emission, and electric and magnetic fields. Japanese and Soviet scientists have been particularly active in observing these phenomena. In 1969, the Soviet scientists found an effect that seems especially important—a premonitory change in ratios of two seismic velocities, the compressional velocity (V_p) and the shear velocity (V_s). The ratio of V_p to V_s changed by about 15 percent in the periods preceding moderate-sized earthquakes in the Garm region of central Asia. American geophysicists attending the 1971 international meeting in Moscow learned of these findings and have now made similar observations in the United States.

American geophysicists have also developed a model to explain the premonitory phenomena. This involves the changes in strength and in velocities of seismic waves in rocks that are related to the presence or absence of water.

The new explanations tie in very well with field observations that have been made in Colorado. The Denver earthquakes of the 1960s were triggered by deep injection of waste fluids. More recently, experimental injections and withdrawals have been conducted in the Rangely Oil Field. These have demonstrated that stresses can be relieved by injections of water that trigger small, harmless events. On withdrawal of fluid, the earthquakes stopped.

These evidences of progress in prediction are important and interesting, but they are only a beginning. The new information seems applicable to shallow earthquakes but may not be relevant to deep events. Moreover, even with the shallow earthquakes, there may be differences in those that are strike-slip and those that represent overthrusting.

Moreover, the very large events could have features that are qualitatively different from the smaller earthquakes, which are readily studied. If we wish to understand and be able to predict the rare, large earthquakes, we should be seeking premonitory signals everywhere that earthquakes have been known to occur. We should invest in new ideas, development of new instrumentation, and in the establishment of observing networks. Other countries should be encouraged to do likewise, and we should assist them whenever feasible.

The task of minimizing earthquake disasters is a large one and may require decades to complete, but what are decades in a span of millions of years?

---

18 July 1980

The eruption at Mount St. Helens brought shock and deep anxiety to the Pacific Northwest and evoked sympathy from others around the world. The eruption was not as powerful as some, but it caused destruction estimated at over $2 billion and the loss of nearly a hundred lives. The death toll would have been much larger had warnings not been issued and had the Forest Service not acted to minimize the number of people close to the mountain.

Geophysicists and petrologists are necessarily somewhat uncertain about the detailed processes that give rise to volcanism. The phenomenon involves magma and temperatures of 1100°C and more. At the pressures and temperature existing at 100 kilometers, it is supposed that partial melting of the rocks there would occur. In areas such as the "ring of fire" that surrounds the Pacific Ocean, the movement of tectonic plates would influence events. Relative movement of the plates would be likely to produce weak areas through which magma might migrate. The magma is less dense than the solid from which it is derived. Rock pressure exerts a powerful force tending to move magma toward the surface.

Evidence from a variety of volcanoes shows that interconnected pools of lava with total volumes of cubic kilometers may accumulate in the quiescent periods between episodes. When such a volume moves toward the surface, seismic activity can be noted. In the two months preceding the eruption at Mount St. Helens, a large number of local earthquakes were recorded. Their number and intensity justified the warnings and the Forest Service restrictions on travel near the mountain.

The seismic activity that began in March 1980 did not come as a complete surprise. On 7 February 1975 *Science* published a report that began, "Mount St. Helens, a prominent but relatively little known volcano in southern Washington . . . has been more active and more violent during the last few thousand years than any other volcano in the conterminous United States. Although dormant since 1857, St. Helens will erupt again, perhaps before the end of this century." Many of the earlier eruptions of Mount St. Helens were dated by carbon-14 techniques. The hot ash converted some of the nearby trees to charcoal. In addition, examination of the geologic column has revealed widespread ashfalls derived from eruptions at St. Helens.

An interesting fact also mentioned in the report is that the composition of the material vented by the mountain on different occasions has varied widely. Correspondingly, the nature of the volcanism has varied. A magma that is high in silica (65 percent or more) is very viscous and tends to be associated with violent eruptions. A magma with about 50 percent silica is much less viscous and flows rather quietly from the volcanic vent. In the past, both types of volcanism have occurred at St. Helens; what form the volcanism will take next is unpredictable. The same is true of other volcanoes, both here and abroad. History tells us that destructive episodes involving great loss of life have occurred and will occur again. We need not go back to Vesuvius and Pompeii. In 1902 Mont Pelée on Martinique commenced to eject ash, which drifted down over the city of Saint Pierre 10 kilometers away. Two weeks later a violent explosion killed all but one of the 30,000 inhabitants.

Most of the Cascade Range volcanoes and those around Cook Inlet in Alaska have manifested volcanic activity of some sort during the last two hundred years. The U.S. Geological Survey proposes to expand its monitoring of them. The program would include emplacing seismometers, tiltmeters, and distance measuring facilities on the more active peaks. In addition, careful studies of earlier ashfalls would be made, including carbon-14 dating to determine the frequency and characteristics of earlier events. Even with detailed knowledge, one cannot hope to forecast precisely when eruptions will occur. However, better knowledge would surely lead to saving many lives. The proposed program, which would cost about $6 million annually, should be speedily authorized, supported, and implemented.

3 July 1981

Few weeks go by when there are not earthquakes or volcanic activity. A great heat engine, fueled at least in part by radioactive decays deep in the earth, relentlessly moves pieces of crust the size of continents. Motions of tectonic plates relative to each other averaging 2 to 10 centimeters per year are common. Such motions persist and their cumulative effects are drastic changes in oceans and land.

New crust is being formed under the oceans at the mid-ocean ridges, and the ages of the oldest rocks on the sea bottom are only about 200 million years. In contrast, the age of the earth is about 4600 million years, and that of the oldest rocks found on the continents is about 3800 million years. Accompanying the movements of the plates have been collisions of landmasses, mountain building, and repeated hot and cold geochemical and biological processing of enormous amounts of material. While most of the continental areas have been tremendously deformed, parts of them have been miraculously left little changed from the distant past.

Employing careful observations and the concepts of uniformitarianism, geologists and paleontologists made great progress in deciphering Earth and biological history. During this century, and especially within the last thirty years, simple observation has been supplemented by tools derived from the physical sciences. In the United States in the 1950s a small minority of earth scientists believed that the continents had moved great distances. It was after geophysical exploration of the oceans, including measurements of magnetization of bottom rocks and deep-sea drilling, that the concept of moving tectonic plates won acceptance.

Analysis has shown that the composition of the magma reaching the earth's surface has changed since earliest times. For this and other reasons, the concept of uniformitarianism must be qualified. How and to what extent are questions to be studied further? Of special interest is the history of the movement of the earth's crust. Earlier, radioactive heat was released at four times the present rate. Surely in the past the earth's engine was more violently active, but in what ways?

To answer questions about events that occurred before 200 million years ago, one must seek testimony preserved on the continents. There one can find evidence of both great lateral and vertical motions and of collisions of tectonic plates. Additional knowledge about the complex structure of the continental crust is being accumulated with new tools. For example, the Consortium for Continental Reflection Profiling has discovered that ancient crystalline rocks of the Appalachian Piedmont and Blue Ridge appear to have been thrust 260 kilometers westward over younger, sedimentary rocks. The data also suggest that the thrusting was related to multiple opening and closing of a proto-Atlantic Ocean.

The current status of geodynamics and opportunities for further research depends on close worldwide cooperation if the past is to be deciphered. At best, evidence is scattered and fragmentary. For example, in this country there are only limited outcrops of early Precambrian rocks. Good occurrences are seen in Greenland, Zimbabwe, and Brazil. Similarly, while California experiences some earthquakes, a better place to study them is Japan, where the frequency is an order of magnitude greater.

Important work was stimulated by the International Geodynamics Project, which ended in July 1980. A new Inter-Union Lithosphere Program has been organized which will emphasize studies of the continents and their margins. Out of this project will come better understanding of the past, better knowledge of earthquakes and allied natural hazards, and an improved basis for discovery of mineral and petroleum concentrations.

1 April 1983

The earth has been the scene of many extinctions during its long history. Some of them have occurred relatively slowly and are readily explained, for example, by gradual large-scale climatic changes or the appearance of successful competitors for ecological niches. But extinctions have occurred that have involved a large fraction of the existing life-forms. Suggestions have been made that such events might have been due to impacts of large bodies from elsewhere in the solar system. However, it was only a few years ago that evidence was presented for the simultaneity of a very large asteroid impact and extinctions at the end of the Cretaceous period (65 million years ago).

The evidence took the form of a very large iridium anomaly in a thin layer in marine sedimentary rocks laid down at the end of the Cretaceous. This work was followed by reports of related occurrences in both marine and nonmarine sedimentary rocks in many different localities around the world. Attention was accordingly focused on questions about the frequency of large-scale impacts and their immediate and longer term signatures. These important questions came to involve efforts by geologists, geochemists, geophysicists, paleontologists, chemists, and physicists and led to a very lively interdisciplinary meeting in Snowbird, Utah, in October 1981. The papers presented at the meeting were recently published in a book edited by L. T. Silver and P. H. Schultz and entitled *Geological Implications of Impacts of Large Asteroids and Comets on the Earth.* It makes good reading for a wide audience.

There are about one thousand asteroidal bodies with diameters greater than 1 kilometer whose orbits cross that of the earth. About three of these hit the earth every million years. Smaller bodies are more abundant and collide more frequently. The postulated Cretaceous projectile had a diameter of about 10 kilometers. Objects of this size are not very abundant and they may strike the earth about once every 40 million years. A 10 kilometers object having a velocity of 25 kilometers per second would bring with it an energy of about 4 x 10^{30} ergs.

Evidence of many collisions is found on the moon, Mars, Mercury, and Earth. On the earth the best-studied impact feature is the Ries crater in West Germany. It is 26 kilometers in diameter and about 800 meters deep and was formed about 15 million years ago by the impact of an object 1 to 2 kilometers in diameter. Studies of the ejecta provide a picture of tremendous manifestations of energy in the form of high pressures, high temperatures, and high-velocity projectiles.

Aside from the iridium anomalies, the principal evidence for a major event 65 million years ago comes from paleontology. Effects differed widely among the various genera on land and sea. Those most affected were planktonic calcareous shelled organisms living in near-surface regions of the tropical oceans. Benthic creatures and siliceous shelled organisms were less affected. John Lewis and colleagues have suggested that a substantial lowering of the pH of surface waters was involved. They point out that, with the high temperatures associated with a large impact, tremendous quantities of nitrogen oxides would be formed. These would be converted to nitrous and nitric acids and would descend to the earth in the form of acidic precipitation. On land the large buffering capacity of soil would neutralize the acid, but at sea the top layer has little buffering capacity and mixes only slowly with deeper waters.

Not all scholars agree that a major impact occurred at the end of the Cretaceous. We are only at the beginning of discovering and interpreting phenomena connected with impacts of large bodies on the earth. The new book provides a valuable benchmark of the state of knowledge and speculations in this important field.

---

12 February 1965

The United States is in the process of committing itself to an expanded space program. The Space Board of the National Academy of Sciences, which has provided many of the goals of the program, recently stated, "The new goal for the period 1971–1985 should be scientific exploration of Mars . . . Mars is of great scientific interest first because it offers the best opportunity in our solar system for shedding light on extraterrestrial life . . . " A search for life on Mars is thus one of the major scientific justifications given for a program that is likely to cost as much as $100 billion during the next two decades.

Our present knowledge of Mars is incomplete, but the facts available provide little basis for thinking that life will be found there. Mars is arid. The total condensable water in a column from the surface of the planet to the top of the atmosphere is about 14 microns (0.00055 inch). White polar caps grow and shrink with the seasons. An average cap thickness of 1 centimeter has been estimated. Considering the small amount to be vaporized and the aridity of the atmosphere, it seems unlikely that liquid water ever exists on the planet. Mars is cold. The average temperature is 230 K. At midday the temperature at the subsolar point can be as high as 298 K, but at night the temperature drops far below freezing, to about 220 K. Recent work indicates a thin atmosphere with a surface pressure of about 25 millibars. In addition to the trace of water, the only constituent known to be present is carbon dioxide (about 5 percent). Oxygen, if present, accounts for no more than about 0.1 percent of the atmosphere. Toxic carbon monoxide, produced by irradiation of carbon dioxide, could be a constituent.

The severity of the Martian environment does not seem to have been realistically taken into account in plans for the exploration of Mars. Exobiologists are very apprehensive lest space probes carry Earth-type organisms to Mars. Extensive and expensive precautions are being taken in an effort to guarantee that there be not one chance in 10^4 of a single organism being carried to the planet. Because of the precautions, many years and billions of dollars could be added to the space program. Before spending large sums on sterilization the space agency should determine whether sterilization is necessary, by encouraging relatively inexpensive studies here on Earth. A few laboratory experiments have been performed in so-called Martian environments. Workers have usually failed to control the water content properly and to test the effects of compounds likely to be produced by solar radiation.

In most proposals for detecting life on Mars it is tacitly assumed that life there would be similar to that on Earth. Some experiments call for the culturing of possible Martian organisms in media brought from Earth. The hypothetical organisms are to be the beneficiaries of an unaccustomed luxurious environment. Two proposed experiments make more sense. In these, gas-liquid chromatography and mass spectrometry would be used. A sample of Martian soil would be pyrolyzed under controlled conditions, and the off-gasses analyzed. By this means compounds suggestive of terrestrial life could be identified. In addition, compounds derived from bizarre forms of life could be observed, as well as unexpected chemicals in the atmosphere and on the surface. In looking for life on Mars we could establish for ourselves the reputation of being the greatest Simple Simons of all time. A few inexpensive experiments to probe the nature of the atmosphere and surface of Mars might save us from considerable eventual disappointment.

10 September 1965

Scientifically, the results of the Mariner IV mission constitute the most important advance in space research since the discovery of the Van Allen radiation belts. Contributing to the value of the mission is the fact that the results of the various experiments are complementary; they also build on and extend previous findings of ground-based astronomy.

Useful data on particles, fields, and micrometeorites were collected during the voyage to Mars. Additional information was gathered after the flyby, and more may be forthcoming when the spacecraft is once again fairly close to Earth. The major contributions, however, are the observations in the vicinity of Mars. Among the most important are the photographs. These show that, unlike Earth, Mars resembles the moon in topography. There are many craters, but there is no evidence of mountain chains.

Recent experiments on particles and fields show other major differences between the two planets. The magnetic field of Mars is not more than 1/1000 that of Earth, and the Red Planet has no radiation belt. An occultation experiment gives independent evidence that the atmosphere of Mars is tenuous and unlike that of Earth. A micrometeorite study shows that interplanetary dust is more abundant in the vicinity of Mars than near Earth.

The evidence from the photographs, the absence of a sizable magnetic field, and the character of the atmosphere all support the view that the history of Mars has been unlike that of Earth.

An example of a close relation between Earth-based findings and findings from the Mariner IV mission is the estimate of the composition and density of the Martian atmosphere. Astronomers have known for some time that the atmosphere of Mars is thin and that it contains carbon dioxide (CO_2). Recently the estimates have been sharpened. Measures of infrared radiation indicate that the total pressure at the Martian surface is 11 millibars, of which about half is CO_2 (0.28 mole per square centimeter). The occultation experiment performed on the Mariner

The Mariner IV Mission

IV mission determined changes in radio signals from the spacecraft caused by passage through the atmosphere and the ionosphere of Mars. Preliminary interpretation of the data provides an estimate of the scale height of the atmosphere ($\sim$ 9 kilometers) and its density. The pressure at the surface of Mars as estimated from the data (about 5 or 6 millibars) is lower than estimates obtained in ground-based studies. This disagreement is not serious, and the discrepancy will probably diminish on further analysis. The important fact is that two very different kinds of measurements give essentially the same result. Half or more of the atmosphere of Mars is CO_2, and the total number of molecules per unit area is about 1/100 the number in the Earth's atmosphere.

The contrast between Earth and Mars can be stated in another way by listing the amounts per unit area of three volatile substances that have appeared at the surface of the planets in the past or are now present. For Earth the values are water (H_2O), 3.2 x 10^5 g; CO_2, 1.8 x 10^4 g; nitrogen (N_2), 8 x 10^2 g. The corresponding values for Mars are: H_2O, $\sim$ 0.01 g; CO_2, $\sim$ 12 g; N_2, $<$10 g. The numbers are not strictly comparable, for most of the CO_2 that has reached the surface of the Earth is now incorporated in sedimentary rocks. Probably most of the H_2O that has appeared on Mars has been lost, the hydrogen having escaped and the oxygen having been consumed or lost. Nitrogen has not been detected on the planet, and the value given is probably an upper limit, derived from the pressure effect it exerts on CO_2.

The success of the Mariner IV mission represents a superb engineering achievement by the Jet Propulsion Laboratory. The accomplishment required the proper functioning of 134,000 parts after seven months in space. The magnitude of the success is highlighted by the failure of others to attain the goal. The Russians, who have some first-class engineering talent, have not succeeded in their dozen or so attempts at attaining close-in data from Mars or Venus.

---

17 March 1967

In support of the administration's recommendations on the future of the space effort, panels of the President's Science Advisory Committee have prepared a report entitled "The Space Program in the Post-Apollo Period." This document seems designed to provide an intellectual justification for a continuing program likely to cost more than $50 billion. In view of the issues involved, one might hope for a comprehensive report delineating and weighing alternatives; the actual product is thin, and it advocates oftener than it weighs.

Major matters that need to be discussed are: What are the major scientific challenges? What is the importance of these questions as compared with those that can be studied on Earth? What are the chances of discovering extraterrestrial life? What are the arguments for manned versus unmanned exploration of space?

In outlining objectives for the post-Apollo period, the report slights near-Earth activities that are likely to pay off well both scientifically and practically. The principal questions set forth are: Does life abide in places other than the Earth and, if so, what is its nature, how did it evolve, and what are its probable forms elsewhere? What is the origin and evolution of the universe, and what is its ultimate destiny? What is the place of our sun and solar system in it? Do natural laws as we know them on Earth indeed govern the behavior of every observable part of the vastness of space? What are the physical conditions on the moon and on the other planets in our system, and how did our solar system evolve? What dynamic relationships between the sun and the planets shape their environments?

These are grand questions, but it was not made evident that the post-Apollo program has much chance of answering more than a few of them. The best prospect for fundamental, scientific findings is a program employing astronomical observatories in orbit.

The report is less than complete in its discussion of the comparative value of space and nonspace activities: "Space programs can be thought of as competitive with other quite different programs, for example, in oceanography, improved transportation, or in urban renewal." However, the difficult problem of priorities was quickly ducked: "Comparisons among the different programs go well beyond the competence of the Panels."

A substantial fraction of the expensive post-Apollo program is to be devoted to a search for extraterrestrial life. However, only a few sentences in the report mention the search. Nowhere is there an evaluation of the chances of finding life on Mars or Venus.

Another deficiency is the lack of a full discussion of the role of man in deep-space exploration. To date, manned missions have contributed little scientifically. The unmanned missions have had a cost-effectiveness for scientific achievements perhaps one hundred times that of the manned flights. Nevertheless, the report implicitly calls for a major role for man in the post-Apollo program.

The advocates of a large continuing space program have made their report. A committee of nonspace scientists would recommend differently. However, they are not likely to be asked to do so.

5 December 1969

This has been a remarkable year for geology and associated disciplines. Two great developments have enriched these fields with new knowledge, new puzzles, and new objects for study. The most widely publicized of these developments is the exploration of the moon. Examinations of lunar samples are progressing well, scientists are very excited about what they have been finding, and some results will soon be released. Investigators have only begun to study materials from the *Apollo 11* landing, and specimens from *Apollo 12* will soon be available.

The second development is the success of an extensive program of drilling of the deep-sea bottom. Holes have been drilled at sixty-six carefully selected sites in the Gulf of Mexico, the Caribbean Sea, and the Atlantic and Pacific oceans. At about twenty of the sites, cores all the way down to the igneous basement have been obtained. To date, examinations of the cores have been conducted on shipboard, but major conclusions have already been announced.

In terms of the total history of the earth, the present ocean basins are relatively recent features. Although the most ancient rocks on the continents are about 3400 million years old, the oldest sediments obtained from the deep-sea bottoms are only 140 million years old.

Results from the drilling strongly support hypotheses of seafloor spreading and continental drift. About 200 million years ago Europe, North America, South America, and Africa were joined, but at that time the separation of Europe and North America began. Examination of cores from the Atlantic Ocean reveals that new crust is forming at the Mid-Atlantic Ridge and is spreading on either side of it. The rate of movement ranges from 1 to 4 centimeters per year. The results confirm earlier views based on magnetic observations. However, the deep-sea drilling has changed speculation into something that must be regarded as established. Geophysicists point out that the drilling has a related benefit. It is fairly easy to survey large areas of the ocean with airborne magnetometers or with ship-carried seismic equipment. Results from drilling now permit confident interpretation of these geophysical observations.

In addition to the lateral movements of the continents there have been vertical motions, both up and down. For example, an area east of Brazil that was once at sea level is now 2000 meters beneath the surface.

Detailed study of the cores will begin shortly. Samples will be broadly available. They will provide an improved history of life, of climatic change, and of geochemical events in the oceans and the sediments.

The National Science Foundation has announced that it will support additional deep-sea drilling, which will facilitate exploration in the Mediterranean Sea, the Indian Ocean, and other as yet untapped areas. In addition, the drilling company Global Marine has plans to develop techniques whereby dulled bits can be changed and drill holes can be reentered. It will then be possible to obtain cores of substantial length in the basement rocks.

One cannot foresee the new knowledge, the new questions, and the new opportunities that will arise from the two great recent developments. It is clear, however, that before another decade is over our understanding of the earth and the solar system will be substantially increased.

10 April 1981

Voyager 1 made its closest approach to Saturn on 12 November 1980. On that day, the mission control center at Jet Propulsion Laboratory was the focus of intense worldwide attention. Interest in the Voyager mission to Saturn approached that accorded the first manned lunar landing.

The Saturn system is a frigid mysterious world nearly 1.6 billion kilometers (1 billion miles) from Earth. Per unit area, it receives about 1 percent as much sunlight as does Earth. Spacecraft visiting it must be prepared to withstand extreme cold, to operate semiautonomously, and to convey messages to and receive messages from Earth.

Saturn has a mass ninety-five times that of Earth. It has an atmosphere that consists mainly of hydrogen, with helium (approximately 11 percent) the next most abundant component. Methane, ammonia, ethane, ethylene, acetylene, and phosphine have also been detected. The temperature decreases from 150 K in the upper atmosphere to a minimum of about 85 K at a pressure of 100 millibars and then increases to about 160 K at 1.4 bars. The planet is obscured by clouds, which move at velocities that are a function of latitude. Eastward wind speeds as high as 480 meters per second (1100 miles per hour) were observed near the equator.

One of the major objectives of the Voyager mission was to gather information about Saturn's satellites. There are fifteen of them, including three that were discovered during the flyby. Titan, the largest of the group, is the second largest satellite in the solar system (Jupiter's Ganymede is first) and the only one known to possess a substantial atmosphere. Although it is covered with clouds and haze, Voyager mission experimenters were able to determine its diameter (5140 kilometers). Using this datum and the mass, they calculated Titan's density to be 1.9, which corresponds to a 50:50 mix of rock and water ice. The atmospheric pressure at the surface of Titan is 1.6 bars and the temperature approximately 93 K. Nitrogen is the main constituent of the atmosphere, with methane next in abundance. At the conditions on the surface of Titan, gaseous, liquid, or solid methane might be present. The other satellites were not obscured by clouds. They were covered with water ice and in some cases are composed mainly of water ice. A striking feature of Mimas is a crater roughly 130 kilometers in diameter. Craters were also observed on most of the other satellites.

Saturn's rings were found to have a far more complex structure than predicted. They consist mainly of water ice. Voyager 1 mission results indicate that the A and C rings contain particles with effective diameters of 10 and 2 meters, respectively. The Cassini division, a classical ring element separating the A and B rings, itself contains five broad rings with substructure. The F ring has an unusual morphology, with two components that appear kinked and braided.

The foregoing paragraphs mention only a fraction of the information now available about the Saturn system. Moreover, only part of the experimental data has been analyzed thus far. When analysis is complete, a very substantial body of facts will be added. For centuries scientists have attempted to answer three major questions about the solar system: How did it originate? How did it evolve? and How does it operate today? The information gathered with manned and unmanned spacecraft greatly limits the range of permissible speculation. A theory that covers the origin and evolution of the solar system will illuminate processes that have occurred on Earth. Data about atmospheric motions on Earth, Mars, Jupiter, and Saturn will be used to test models of global circulation.

The Voyager 1 mission to Saturn has been another great success in a long series of U.S. exploits in space. The engineers, scientists, and technicians involved in the era of space exploration can take pride in their work. They have participated in one of humanity's greatest achievements.

---

23 May 1980

Major industries with total annual sales of over $500 billion are intensively engaged in the development of new and better materials. Their efforts are crucial to innovations that will render this nation more energy-efficient and more capable of meeting international competition in the future. The internal atmosphere of the best R&D laboratories is favorable to the speedy translation of research results into applications. Because the companies recognize that their future depends on research, the scientists involved enjoy excellent support and are provided with abundant facilities. Some of their equipment defines the state of the art. In many areas of science pertinent to materials, industrial scientists are the pioneers.

Three general groups of materials are involved: polymers; metals, alloys, oxides, and silicates; and electronic materials, primarily semiconductors. New kinds of polymers continue to be discovered having special properties such as great strength, high thermal and chemical stability, or electrical conductivity. Fundamental understanding of the behavior of polymers is being obtained by use of nuclear magnetic resonance and other experimental tools, which give quantitative guidance in efforts to formulate superior products. Major activity, though, is devoted to combining already available monomers and polymers to form objects with desired properties superior to those of a pure polymer. For example, combinations of layers of polymers can lead to containers that are tough, strong, and resistant to passage of oxygen. Mechanical properties can be greatly altered by incorporation of reinforcing fibers, inert materials, or gases. The new products are finding many uses in energy-saving applications.

In their studies of materials, chemists and physicists have roamed throughout the periodic table and have made countless combinations of elements and tested them in various proportions. Such work has led to new superconductors and to improved permanent magnets that require less imported cobalt

Advanced Technology Materials

than earlier types. Major advances are being made in improving the strength of materials. One method takes advantage of the fact that some crystals have great unidirectional strength. Another development is the creation of low-alloy, high-strength steels. Even more spectacular has been the development of glassy metals. When liquid mixtures are cooled very rapidly, the resultant solids may have strengths fifteen times that of products cooled more slowly. At the same time, other properties such as magnetic permeability and freedom from corrosion may also be greatly improved. A research effort of great importance is the work to develop superior specific catalysts. This involves detailed understanding of the interactions among atoms at surfaces. Improvements of as much as a factor of 10^{12} have been obtained in speeds of reaction. When combined with high specificity, such performance leads to major energy savings. The research effort on catalysts has also led to the development of zeolite cage structures capable of catalyzing the conversion of methanol to gasoline.

During the past decade the most dynamic area of technology has been in exploitation of the potential of semiconductors such as silicon. The electronics revolution continues with considerable emphasis on obtaining more transistors per chip and better, lower cost computer memories. But other frontiers are under scrutiny. Semiconductors such as GaAs (III-V compounds) may be the key to even faster, better computers. Such compounds have already proved useful for lasers and light-emitting diodes. A different approach to increasing the speed of computation is through the development of Josephson-type devices that function at cryogenic temperatures. Another activity is work to develop superior photovoltaic materials.

One of the fastest growing applications of new materials is in prosthetic devices. This year, between 2 and 3 million such devices will be implanted in humans, creating an interesting set of interactions between living and nonliving substances.

---

9 November 1984

Ability to compete in high technology will be an important determinant of success in the competition between nations and regions. In turn, success in technology is crucially linked to leadership in materials science. The development of new or improved materials permits the creation of new technologies, better quality control, less costly products, and, in some instances, less dependence on imports of scarce elements. To advance materials science requires close collaboration of physicists, chemists, and engineers. Laboratory experimentation is stimulated by theoretical physicists or chemists. Scientific advances also come as a result of stimuli generated by engineering needs.

For the most part, the chemistry of the substances available for use as materials is well known. However, the behavior of substances at surfaces and interfaces is not well understood. Microstructures, both in the interior and at the surface, can have a large effect on properties. One method of markedly altering a surface is ion implantation, which is widely used in the semiconductor industry. It is also increasingly being used in other applications to provide surfaces that differ radically in composition from the bulk of an object. Intense, but short-period, laser irradiation can almost instantaneously melt a shallow layer at a surface; this is followed by a comparably fast solidification. This treatment can freeze into place microstructures with superior resistance to friction, wear, and corrosion. Laser treatment of engine parts is now widely used in the automobile industry.

Research on the surficial interactions between living tissue and prosthetic devices has led to potentially improved performance of the millions of implants that are done each year. Particularly useful are glasses or ceramics containing some calcium phosphate. At the surface of the prosthetic device in the presence of body fluids, hydroxyapatite is formed. Cells recognize this substance, are compatible with it, and form connections to it.

In many applications, engineers seek materials that maintain their tensile strength at high temperatures and that are resistant to corrosion. Many of the alloys now in use contain imported chromium as an essential constituent. The intermetallic alloy Ni_3Al is ordinary brittle and unworkable. However, with the addition of boron (200 parts per million), it becomes ductile. Its strength actually increases with temperatures up to 900°C, and it is comparatively free from corrosion.

Plastic composites seem destined for very large scale applications in the aircraft and automobile industries. Already one manufacturer of airplanes has announced a new model to consist largely of plastics. There will be great savings in weight as well as fewer parts to assemble. A large fraction of the Pontiac Fiero will consist of composites. Intrinsically, some polymer molecules are extremely strong. To use them advantageously requires a knowledge of their behavior during processing. Once strong fibers are formed, composites may be created to meet engineering specifications.

The impetus for theoretical studies and research aimed at revolutionizing computers and data transmission comes from the successful development of optical means of communication with the use of glass fibers. It also is based on the development of lasers that are capable of delivering extremely short pulses. In principle, computers based on photon processing could operate much faster than the current semiconductor devices.

Efforts to advance materials science will continue to be driven by opportunities and international competition. Students who have an aptitude for the physical sciences and a desire to participate in research with obvious payoffs in meeting societal needs will find work in the materials sciences rewarding.

2 April 1971

Cancer is an enemy of all mankind, and whoever helps to attenuate its effects will be the benefactor of untold billions of people. With so much at stake the public would like to believe that a massive centrally directed program could conquer the disease. Unfortunately, a sharply focused effort has no guarantee of success and could damage other health research programs.

The basic nature and origins of cancer are complex. Countless related diseases are subsumed under the name of cancer with numberless causes, different courses, and different prognoses. For example, it has been demonstrated that more than one hundred viruses and more than one thousand chemical substances can produce cancer in animals.

A major recent source of information on the status of cancer research is the report of the National Panel of Consultants on the Conquest of Cancer for the U.S. Senate. The report devotes much attention to results of laboratory studies on the disease. These are interesting and hold much promise. However, the report also points out, "Cancer prevention offers greater possibilities for the control of cancer and the saving of human lives than any other measure now at hand. Many, perhaps most, human cancers can now be prevented. The most important environmental causal agent in the production of internal cancer today is the prolonged inhalation of cigarette smoke." Other preventable cancers include those due to environmental agents such as arsenic, asbestos, coal tar, and radiation. Fragmentary evidence indicates the heavy involvement of yet unidentified environmental and social factors that may be preventable. Primary cancer of the liver may be as much as five hundred times more frequent in the African population of Mozambique than among black people in the United States. The probable cause is aflatoxin, a carcinogen produced by a mold that can grow on peanuts or cereal grains. Cancer of the colon and rectum is the leading internal cancer in the United States but is infrequent in Mexico and Latin America. American women have about seven times more cancer of the breast than Japanese women. Incidence of stomach cancer is high in some countries but has been decreasing in the United States. Sharp discrepancies in cancer incidence seem related to environmental, not genetic, factors since they hold true for people with similar genetic constitutions in different environments.

Next to prevention, the most effective approach to cancer seems to be early detection and treatment. Statistics show that the prognosis deteriorates badly as the cancer becomes more advanced. Cytological tests, notably the one developed by Papanicolaou, have been very helpful in early detection of cancer, leading to successful treatment. Chemical and immunological tests for early detection seem promising, and they will certainly be developed further. Very tantalizing but still unevaluated are hopes that immunological knowledge and techniques may prove helpful in the treatment of cancer.

In view of the complexity of the diseases known as cancer, we cannot reasonably hope for a magical single cure. The time that will be required for substantial attenuation of the effects of cancer is decades. Increase in long-term support for cancer research is fully justifiable and should be implemented. However, the likely result of a hurry-up-and-wait crash program is wreckage of the nation's medical research enterprise without much counterbalancing progress in coping with cancer.

18 June 1971

A major determinant of the quality of future civilization will be the wisdom and effectiveness with which man deals with renewable resources and with the natural environment. Central to good management of these matters is first-class competence in the plant sciences. Recently there has been much talk about ecology and the environment, but there has been no corresponding acceleration in the undergirding fundamental science.

At one time botany and zoology were roughly coequal in biology at universities. The emergence of large federal support for medically oriented research changed that relationship. Some aspects of botany, such as growth, were supported moderately by the National Institutes of Health, as was photobiology, including photosynthesis. Other aspects, such as ecology, were not encouraged. Thus, botany came to be overshadowed in some universities and lost identity and stature.

The financial strains of the past few years have been felt rather keenly by plant biologists. The NIH and the Atomic Energy Commission have found it necessary to diminish their support. The National Science Foundation has begun to increase its funding for botany, but the level is still quite low.

Botanists are almost unanimous in their disappointment that the Agricultural Research Service (ARS) has not chosen to institute a grant system comparable to that of NIH. Although academic botanists concede that agricultural research has been cost-effective, they feel that ARS has not given sufficient support to work of a truly fundamental nature.

Given an improved intellectual climate and a moderate increase in funds, the plant sciences would flourish. There are substantial matters, both applied and fundamental, to address. The practical challenges facing plant biology include applications in temperate and tropical agriculture and in management of fields and forests. We have developed extraordinarily productive farm crops, but monoculture and the use of limited strains of plants makes the food supply vulnerable to plant enemies such as the southern corn leaf blight. Most of our agricultural research has been devoted to plants of the temperate zone, and the knowledge acquired is not readily adaptable to tropical conditions. Success to date of the Green Revolution indicates what might be accomplished.

A superb group of tools and techniques developed for use in animal biochemistry can be employed effectively in the study of plants. As one example, the use of amino acid analyzers has been crucial in the selection of maize mutants possessing a high lysine content and correspondingly high nutritive value. Recently, it has become clear that plants are involved in complex chemical warfare with pests and with each other. Greater knowledge of the biochemistry of plants will add an important new dimension to comprehension of ecological relationships. The use of atomic absorption equipment can enlighten us on requirements and utilization of limiting trace elements. One of the developments that seems particularly useful is the creation of mobile laboratories, which enables investigators to study the behavior of plants under a wide range of natural conditions. Thus, the performance of a twig or leaf can be measured under controlled conditions while still attached to a plant.

Research opportunities in many aspects of botany await the energetic and imaginative investigator. Modest increases in support for fundamental research in the plant sciences would bring beneficial returns of disproportionately large magnitude.

22 June 1984

To those whose experience in chemical research laboratories was twenty years ago or more, the modern counterpart is a strange place. New instrumentation with electronic components has revolutionized analytical capabilities. It has also made accessible crucial experiments and theoretical calculations that could not previously be performed. Some present-day measurements can be made with speeds and sensitivities that are five to ten orders of magnitude better than those of two decades ago.

In a symposium at the recent AAAS meeting in New York and in an earlier report [*Research Briefing Panel on Selected Opportunities in Chemistry* (National Academy Press, 1983)], leading chemists were enthusiastic about new opportunities for research that have been created. They emphasized three major frontiers. The first is the opportunity to understand, in the most fundamental sense, chemical reactivity and how to control it. The second is to improve understanding of catalyses. The third is to extend to the molecular level understanding of life processes. A few examples are appropriate.

The study of why and how chemical changes take place has been especially facilitated by new instrumentation. Lasers, computers, molecular beams, ion cyclotron resonance, and many more tools have opened and facilitated new research approaches. Of these, lasers have been particularly helpful. Their short pulse durations permit probing of chemical reactions in times ranging from 10^{-6} to 10^{-12} seconds. Lasers also provide tunable, extremely narrow frequency light sources and thus greater diagnostic sensitivity and selectivity. With high-power sharply tunable lasers it is possible to excite one particular degree of freedom of many molecules in a sample. During the interval in which these excited states persist, such molecules react as if that particular degree of freedom is at a high temperature while all the rest of the degrees of freedom of the molecule are cold. Today we know much about the chemistry of molecules at the ground state. The study of their behavior under excitation

Chemistry Without Test Tubes

will greatly improve our understanding and ability to devise important applications.

Catalysts are already important technologically. It is estimated that 20 percent of the gross national product is generated through their use. Much of the present art was developed through empirical research. New equipment facilitating fundamental studies of processes on a molecular level is now available. One result is the rapid development of surface science. Because of the unsatisfied bonding capability of atoms at surfaces, the chemistry there is different from that of reactants brought together in solution or as gases. When chemists are able to identify molecular structures on the surfaces, they will be able to understand and control events there.

Other frontiers of research include homogeneous catalyses, metal cluster chemistry, and stereoselective catalysts. An important branch of homogeneous catalysis has developed from research in organometallic chemistry. An example is rhodium dicarbonyl diiodide employed in the commercial production of acetic acid from methanol and carbon monoxide. Stereoselective catalysts now being discovered will surely have important applications in the synthesis of biological molecules. If a complex molecule has many chiral carbon atoms and a synthetic process produces all of them, only a tiny fraction of the product is likely to have the desired biological activity.

The chemists state, and rightly so, that their science has been underfunded relative to other major disciplines. They point to many new research opportunities and to the needs of the $175-billion chemical industry for new knowledge and trained people. They remind us that although chemicals now produce a favorable balance of payments of $12 billion a year, leadership in some areas of research has moved to other countries. Their pleas for funds to enable them to purchase state of the art instrumentation should be granted, and their efforts to support training of the next generation of chemists supported.

APPLICATIONS

Philip Abelson's interest in the world of science has not been confined to what happened in the laboratories and at the frontiers of science. His view in his editorials also focused outward, on the world and the issues that concern society. And to those issues he brought the generally clear-eyed vision that a thorough grounding in the scientific outlook can sometimes give, together with remarkable common sense and a gift for penetrating to the core of the matter. Thus the editorials collected here, representing that outward vision, are concerned with the applications of science in the world at large and with how this information was used or misused or ignored by society. They include such subjects as the rise of China as a military power, deaths from heroin, competition among nations, and such technological frontiers as the transformation of communications through optical fibers.

Well before jogging became a fad, for example, Abelson pointed out the advantages—mental and physical—of aerobic exercise. Well before industrial policy became a campaign issue, he wrote about the chemical industry and the electronics industry as foundations of our prosperity, noting this country's vanishing lead and the importance of federal policies in maintaining our competitiveness.

He has written about world food supplies, about the continuing improvement of grain crops that have helped many developing countries keep pace with expanding populations, about the potential for farming in the tropical rain forests. He has called attention to malnutrition and its effects on health and mental development—in this country and abroad. And he has written about new approaches in agriculture such as methods to control pests that minimize use of toxic chemicals or efforts to desalinate water.

The applications of science to medicine has been a recurring theme. Abelson has written repeatedly about the risks of cancer from man-made and naturally occurring toxic chemicals in our environment and especially in our diet. Yet he has been equally concerned to combat overreaction to fears of environmental carcinogens, pointing out that smoking and the food we eat account for most of our exposure to toxic chemicals. He has written with equal balance and common sense about the connection between the general health of the population and federal health expenditures, foreshadowing the personal health boom by pointing out that major improvements in health depend more on individual life-styles than on medical breakthroughs.

Abelson has also been interested in the social impact of new technologies. The examples gathered here include concern about the misuse of the power of the computer to invade personal privacy, attention to the widespread utility of electronic databases, acknowledgment of the legitimate social anxiety over rapidly advancing biomedical technologies, and dismissal of the unrealistic, extreme fears of genetic engineering.

30 October 1964

The first official Washington comment on the significance of the recent Chinese detonation correctly indicates that new short-term hazards have not greatly increased, but it does not adequately recognize some longer-term problems.

Only a few facts are available to help one in evaluating the significance of the Chinese detonation. The Atomic Energy Commission has stated, "Additional evidence on the Chinese Communist test of October 16 indicates it was a fission device employing U^{235} . . . ," and, "U.S. intelligence has always led us to estimate that the Chinese Communists were constructing both plutonium production reactors and gaseous diffusion isotope separation facilities."

Production of weapons-grade uranium-235 is an impressive technological achievement indicative of considerable industrial capability. Successful construction and operation of a gaseous diffusion plant capable of producing substantial quantities of weapons-grade U^{235} requires both matériel and skill. Such a plant contains more than a thousand individual units connected in series. Each unit must be constructed with precision—small imperfections can destroy their effectiveness. Moreover, special metallurgical techniques must be available. After the units are assembled their performance must be monitored, controlled, and integrated. This requires a great deal of electronic instrumentation.

A technically incompetent people could not have succeeded in producing weapons-grade U^{235} without massive help; the French, after six years, have not yet announced production of highly enriched uranium. Nevertheless, the new accomplishment was not surprising to many U.S. scientists who have had contact with individuals of Chinese extraction and have known of their first-class aptitude for science and technology.

When a nation builds a successful gaseous diffusion plant it gains great flexibility in nuclear technology. A plant that can produce weapons-grade U^{235} can be tapped to yield uranium having almost any U^{235} content. In this country our nuclear power reactors often utilize material containing U^{235} in the range of 1.5 to 4 percent. The Chinese have the option of producing such uranium. Problems of constructing a reactor are greatly simplified when enriched uranium is available. When ordinary uranium is used, together with graphite, great care must be taken to avoid loss of neutrons either to nuclear poisons or through escape from the reactor to the shielding. With enriched uranium, reactors may be smaller and a wider variety of construction materials can be used. Thus, with enriched uranium, the Chinese have available more options in designing reactors for efficient plutonium production or other purposes than they would otherwise have.

More serious is a greatly enhanced capability of producing tritium, a key constituent of thermonuclear bombs. Tritium is often produced by the reaction of neutrons with lithium-6. Introduction of lithium into an ordinary reactor tends to stop the chain reaction. This tendency can be overcome by introducing enriched uranium. If the Chinese do not now possess quantities of tritium, they can obtain it. In view of the Chinese achievement thus far there is no basis for hoping that they will not achieve a hydrogen bomb—perhaps in the latter part of this decade.

Another member has joined the nuclear club. He already has impressive credentials, and his long-term potentialities should not be underestimated.

27 September 1968

Despite advances in medical research and practice, life expectancy in the United States is not changing much. Benefits of improved practice are being counterbalanced by effects of deleterious living patterns. Heart disease, the number one killer, is increasing. Contributing heavily are such controllable factors as sedentary living, obesity, and excessive cigarette smoking. The Framingham [Massachusetts] Study has identified many factors contributing to heart disease. In one of its contributions it examined the role of physical activity. The most sedentary individuals had a mortality from coronary heart disease five times that of individuals who were active. C. W. Frank and his colleagues in New York have studied coronary heart disease occurring in a defined population of 110,000 men and women. They have noted that the incidence of rapidly fatal myocardial infarction among sedentary individuals who are smokers is nine times the incidence among physically active nonsmokers of similar age.

Physicians have long recommended physical fitness, including weight control, but the admonitions usually were not accompanied by realistic prescriptions. A physiological way to reduce weight is to eat a balanced diet representing a modest caloric deficit with respect to ordinary needs and attain a substantial caloric deficit through exercise. By this means weight losses of 1 or 2 pounds a week can be achieved comfortably. Two recently issued paperback books [W. J. Bowerman and W. E. Harris, *Jogging* (Grosset & Dunlap, 1967) and K. H. Cooper, *Aerobics* (Grosset & Dunlap, 1968)] give simple, effective programs of exercise for improving physical fitness. These books emphasize the great importance of conditioning the cardiovascular system. The value of isometrics is questioned, and calisthenics are assigned a secondary role.

The cost, in time and effort, of achieving fitness is small. Not more than an hour or two a week of effort is required, split among three or more occasions. In sixteen weeks or less, sedentary individuals can note profound changes in their physical condition. For example, the resting pulse rate may be lowered from 85 to 65 beats per minute, and the time required to go a mile may be cut from 12 minutes to about 8.

Two cautionary bits of advice seem indicated. It is well to have a physical checkup before expanding one's activity. In the conditioning process one should "train, not strain." The programs begin with easy tasks—for example, 1 mile of walking—and then progress to more exacting efforts as fitness improves. The recommended means for maintaining fitness is jogging or running. However, other activities, such as cycling or swimming, also are effective.

As one result of the conditioning process, the heart becomes much more efficient. P. O. Astrand has remarked that, by expending a "total of some 2000 extra heart beats during a day's training, you save 10 to 30,000 beats over the remainder of the day."

An important potential benefit from physical fitness is an improvement in mental attitude. The typical middle-aged sedentary individual is in effect crawling submissively toward the grave. By investing an hour or two a week, he or she may add years to a useful life and become a better athlete than many who are twenty to thirty years younger. Buoyed by pride and a sense of well-being, the physically fit individual can compete more confidently in intellectual areas.

17 March 1972

The public and its leaders are now aware of and rightly concerned about the unpleasant potential consequences of overexuberant growth. Small wonder, then, that a recent one-day symposium on the "Limits to Growth" at the Woodrow Wilson Center in Washington should draw an attentive audience that included senators, ambassadors, and a cabinet officer, as well as numerous representatives of the press, radio, and television.

The principal speaker was Dennis Meadows, who has headed a study group at Massachusetts Institute of Technology. This group has used high-speed computers in an attempt to examine the interacting consequences of continued exponential growth of population, food production, industrialization, pollution, and consumption of nonrenewable natural resources. Meadows' major conclusion is the perhaps obvious one that, if mankind continues to indulge at current rates in various forms of exponential growth, calamities will occur in about fifty years.

As a pioneering effort to evaluate a complex web of interactions, the study, and a popular book based on it, is likely to have considerable impact. The book [Meadows et al., *The Limits to Growth* (Universe, 1972)] is to be translated into half a dozen languages and distributed broadly. The approach, though, has defects.

Responding to the allure of tackling a truly big problem, the study group has attempted to treat the entire earth as a single system. In order to limit the number of variables, they chose to aggregate variables on a global basis. Thus the population growth of all the world is taken as a single variable, although the growth rates in various countries and regions differ by a factor of 6 or greater. An even less defensible aggregation of variables is subsumed under pollution. Here an attempt is made to lump together the many components of solid, liquid, and gaseous pollution, even though each differs in nuisance or toxic characteristics.

Enthusiasts can easily lose sight of the limitations of computers. In this study, data from the past are used as a basis for predicting the future, but such data are far more relevant to the past than to the future. The computer is unaware of changing human behavior. Concern about the population explosion and such matters as pollution has already had profound effects. Recently released figures show that the rate of growth of population in this country has been slowing. The rate of increase of important polluting chemicals has been slowed, stopped, or is even decreasing. Important segments of the middle and upper-middle classes are talking of a future "no growth" society.

The study also does not adequately take into account ingenuity with respect to natural resources. Current technology is based on the availability of certain raw materials, such as copper and natural gas. As these resources become scarce, technology will change and, for example, aluminum will be used as a conductor, while methane will be obtained from coal.

The concept of Earth as a closed system is an appealing one, and in some respects it is valid. We all share the oxygen of the atmosphere and must be concerned with changes in its carbon dioxide content. However, much of pollution is local or regional in its effects. The same is true with respect to population. Growth of population exacerbated by concentration in urban centers is a bad enough problem in the United States. It is a far more serious and potentially catastrophic phenomenon in the so-called lesser-developed countries.

A member of the audience at the recent symposium privately reminded us that, although Meadows predicts hell in fifty years, hell is already present on Earth in places such as Calcutta.

9 November 1973

Most natural scientists would readily agree that research in their various fields owes much to tools and insights derived from chemistry. However, fewer realize the full extent of its contributions. For example, biological and medical scientists, while heavily engaged in applying chemistry in their problems, are not aware of its crucial roles in solid-state physics. Lack of awareness extends to an even larger sphere—the pervasiveness of chemistry in efforts to meet societal needs. True, we have all noted applications such as plastics or pharmaceuticals, but we have not had the opportunity to see a comprehensive summary of the great number of ways in which our daily lives are conditioned by products and knowledge that have come out of research laboratories.

Of all the sciences, chemistry has been about the most ineffective in its public relations. This has not been owing to some defect in the character of chemists or their leadership, but rather to special circumstances. Chemistry does not frequently give rise to "spectaculars." Moreover, most chemists work for industry, and companies usually prefer to maintain a low profile.

A recently issued six-hundred page report, *Chemistry in the Economy* from the American Chemical Society, maintains the tradition of a low-key approach, but it does set forth an admirable summary of the role of chemistry in meeting human needs for food, clothing, shelter, health, energy, communications, transportation, and so forth.

The heart of the report is a series of chapters dealing with such topics as food processing, textile fibers, and electronic equipment. Each of their chapters was prepared by a panel of experts drawn mainly from industry. In each case, an historical approach is taken, with key innovators identified. Major products and processes are explained with sufficient detail to be informative, while not excessively technical. At no point does the report talk down to the reader, yet a university student in freshman chemistry could understand it.

As with most such reports in the various disciplines, it

is not entirely clear why the report was written or who was the intended audience. No matter. The report should be read by all academic chemists and their students. The quality of the substantive material is such that it could be used as part of a course. Most chemistry graduates are employed by industry. They and their professors would function more effectively if all understood more clearly what the students were preparing themselves for. In addition, scientists in general who read the report would broaden and enrich their comprehension of the science behind their daily lives.

The report has a particular timeliness. It has been issued at a moment when we must begin to prepare for major changes in the shape of our economy. Our present standard of living is heavily dependent on petroleum hydrocarbons, both for energy and chemical feedstocks. Our economy is also based on the use of a wide variety of other raw materials—many of them imported. The coming years will bring global shortages and high prices, and we will be under strong pressure to make do with domestic resources of energy and materials. Changes in processes and the creation of vast new chemical complexes will be necessary to meet the evolving realities. As the enormous extent of the changes and of society's dependence on science become more apparent, chemistry will emerge as the star performer.

12 September 1969

Two decades ago, American prosperity was solidly based. We had bountiful resources; our industrial plants were undamaged by war; we led in mass production techniques; and our innovative scientific capabilities were outstanding.

Today the foundations of American prosperity have crumbled to an extent not generally recognized. We still possess great natural resources, but they are not adequate to maintain a high-level economy. We face the necessity of importing more and more raw materials and finding the means to pay for them. This will be increasingly difficult, for our ability to compete in international trade is diminishing. In 1964—a good year—U.S. exports exceeded imports by $7.1 billion. In contrast, during the first half of 1969 the value of exports topped that of imports by only $0.15 billion.

An even greater factor than increasing imports of raw materials has been the invasion of foreign finished products such as steel and automobiles from countries that have more than recovered from the destruction of World War II. Our advantage of leadership in mass production techniques has largely disappeared. We still lead in scientific research and in the ability to innovate, but we have lost momentum.

A large contributor to our present problems has been the steel industry. Today, in spite of advantages in raw materials, it does not compete with the steel industries of Germany and Japan. It has been complacent, and slow to adopt the basic oxygen furnace.

In contrast, our chemical industry has long been a leader in research activity. Thus it comes as an especially painful blow to learn that the U.S. chemical industry, which has contributed much to our balance of payments, is feeling the effects of severe foreign competition. This fact was documented in an article by J. G. Tewksbury in a recent issue of *Chemical and Engineering News.* He cited as an example one of the crucial petrochemical intermediates, ethylene. This substance enters into plastics such

Crumbling Foundations of Prosperity

as polystyrene and polyethylene and also into other key chemicals. Analysis of the production and distribution of major items based on ethylene reveals a dramatic change in the U.S. position. Five years ago the United States synthesized about 95 percent of the ethylene products entering foreign trade. By last year this figure had dropped to 40 percent. The big new factors were the European Economic Community and Japan, both of which changed from net importers to heavy exporters. Additional plants are being constructed in Europe and in Japan, and it is quite likely that a further diminution of the U.S. export status will occur.

Tewksbury notes that a number of factors help account for the loss of our competitive ability. The European nations and Japan encourage exports and discourage imports more vigorously than we do. Domestic producers of petrochemicals are handicapped by the oil import program that inflates the cost of their feedstocks. Another factor is the high cost of labor. A few years ago, such disadvantages for the United States were more than counterbalanced by larger plants and advanced technology. These advantages have disappeared. Plants abroad are now of the same scale, and our technology has been disseminated.

The loss of a competitive edge in this area of the chemical industry is a very serious development. It portends similar changes in other areas of high technology.

5 December 1980

America's scientists, engineers, and entrepreneurs made superb contributions to the electronics revolution, and achieved international commercial leadership in many applications such as television, computers, and communication. Our role in TV production has long since atrophied. Until recently, our status in microelectronics seemed unassailable. But our position is eroding.

Two major factors have been involved in our loss of a clear lead: contrasting policies of the governments of the United States and Japan, and Japanese skills in quality control and robotics. Some years ago the Japanese government determined that a strong position in microelectronics was important for the nation's future. A special fund of several hundred million dollars was made available and the various electronic companies were encouraged to cooperate in advancing the state of the art. (In the United States such cooperation would make the companies targets of the Justice Department.) The companies were also accorded fast depreciation treatment of capital costs for new and expanded production facilities. In a fast-moving field such as microelectronics, production equipment becomes obsolete in a short time. The U.S. government took little heed of what was happening, made no realistic changes in tax policy, and instead persisted in endless, bruising, and enervating antitrust suits against two of our best innovators in electronics—the International Business Machines Company and the American Telephone and Telegraph Company.

Because of financial considerations related to slow depreciation of capital costs, U.S. makers of 16K random access memories (RAMs) did not expand their capacity rapidly enough to satisfy the market in 1977 to 1979. The Japanese moved quickly to fill the gap and now have about 40 percent of the market for 16K RAMs, key components in computers. They did more than produce chips. They produced chips of superior quality.

Testimony in this matter has been provided by Richard W. Anderson, manager of the Data Systems Division of Hewlett-

America's Vanishing Lead in Electronics

Packard Company. Hewlett-Packard is the world's largest manufacturer of electronic instruments and one of the three largest users of semiconducter memory in the United States. In 1977, when a shortage of domestic 16K RAMs occurred, they were obliged to turn to Japanese versions. After very rigorous tests, they found that the Japanese 16K RAMs had fewer inspection failures, fewer failures in production cycles, fewer failures in customers' hands, lower rework costs, fewer production interruptions, and lower warranty costs.

I had an opportunity to see something of Japanese quality control at the Matsushita color TV plant near Osaka. At the plant, components and products are subjected to accelerated aging tests at extreme temperatures and humidities. Matsushita makes all of its own components and can assure itself that it assembles only parts that are reliable. About 80 percent of the parts for a color TV set are put in place by robots. Their use cut a rejection rate of 5 percent to a small fraction of 1 percent. At the end of the tour, I realized that I had seen only one inspection station—at the end of the line. I asked the official accompanying me why there were no inspection stations at intermediate points. He replied that until a few months ago there were such stations, but they never found any defects and so they were scrapped. When I returned to the United States, I tried to arrange to see a comparable plant here. I was told that none existed.

Visits to other facilities and laboratories in Japan left me with the impression that the Japanese are not supertechnologists. They merely do what is necessary to build things very well. They will be even tougher competitors in the future.

24 January 1975

For decades the United States attempted to cope with an agricultural system that produced far more food than could be consumed or sold. But in the course of only two years, food stockpiles disappeared and prices soared. Relative complacency has been replaced by anxiety about the world's ability to feed its growing population. This concern has been fed by sympathy for those suffering acute hardships, as in the Sahel and in Bangladesh. It has also been fed by increases in prices at the supermarkets. Many now conclude that the world has reached the limit of its ability to feed even its present numbers adequately. Obviously, if the present birth rates are maintained eventual large-scale tragedy is inevitable. The question is when and where. A recently issued fact-filled report, *The World Food Situation and Prospects to 1985* from the U.S. Department of Agriculture, does not attempt to answer these questions directly, but it provides a wealth of information about the problems.

One of the puzzling features of the present situation is that it has arisen despite a steady increase in global food production. Thus in the period 1954 to 1973, world production grew by 69 percent and, despite substantial population growth, per capita production climbed by 17 percent. In the developed countries per capita production jumped by 33 percent. The picture was not so favorable in the developing countries, where production increased by 75 percent but per capita output climbed only 8 percent. Among these countries great disparities exist, and some regions have fared better than others. For example, during much of the period 1954 to 1973 per capita production of Africa trended downward while that of Latin America was moving up. Some countries, such as the oil exporters, are rich and can easily afford to import food, and others, although poor, are surplus producers. The problems lie with those who tend to be chronically short of food, and are poor.

Further variations are encountered when the prospects for expansion of product are examined. For the world as a whole,

The World's Disparate Food Supplies

45 percent of the land is suitable for crops but only 26 percent is used. In the Asian and Far East region, 84 percent of the suitable land is in crops, while in Latin America only 23 percent is being so used. Thus, there is little room for expansion of the farming area in Bangladesh and India, but a large potential in this hemisphere. The possibility of expanded production of food exists virtually everywhere through use of improved seeds and fertilizers. This is especially true of the developing countries, where yields today are only about half of the developed countries.

Thus it appears that despite a growing population the world is nowhere close to universal famine. However, there are great differences in both current production per capita and future potentials. Unless the rich countries provide a combination of food, fertilizer, and technical assistance, some of the poorer countries face repeated famines arising, for example, from unfavorable weather.

One of the proposals that should be implemented from the recent Rome Conference is the creation of a food reserve to help meet fluctuations in supplies. Existence of such a stockpile would facilitate making lifesaving shipments. It would also indirectly benefit almost all the world's peoples. Because of inelasticity in demand for food, small changes in the balance of supply and demand produce very large fluctuations in price, which could be smoothed out if a suitable food reserve existed. The annual cost for maintaining an effective global reserve has been estimated at only about $550 million to $800 million.

16 April 1982

It was a sunny day in late March on a field of irrigated land located in the Sonora desert of northwest Mexico. The wheat stems were heavy with grain. They would be harvested in about two weeks. On closer examination, it was evident that the field of several hundred hectares was subdivided into thousands of experimental plots. Thousands of varieties of wheat were being tested, some for yield, others for resistance to disease.

This beautiful scene was viewed by an international group of visitors during Presentation Week at the Centro Internacional de Mejoramiento de Maíz y Trigo (CIMMYT). This is the organization whose progenitors developed the wheats that played a major role in the Green Revolution. Today, CIMMYT is a nerve center of an international network of scientists located in more than one hundred countries. When superior varieties of wheat are produced either in Mexico or elsewhere, they are tested in as many as 250 locations around the world. When a new mutant form of a pathogen is detected within the cooperating countries, the news is spread and seed from resistant varieties is multiplied. The varieties of wheat being grown in Sonora represent only a small fraction of the germ plasm that is available. In reserve in wheat banks are more than ninety thousand entries.

To facilitate its work, CIMMYT has several stations located in different climatic zones of Mexico. It is possible to expedite development of new lines of wheat by obtaining two harvests per year, one at Sonora and another in the highlands near Mexico City. The existence of stations in diverse environments also facilitates development of lines capable of thriving under comparatively adverse conditions.

Other programs designed to foster better yields from marginal lands include a form of genetic engineering. That is, wheat is being crossed with wild grasses that can grow under very unfavorable conditions, such as in seawater. Earlier, in Manitoba, a cross of rye and wheat was made called triticale. Varieties of triticale now yield as well as or better than the best wheat and they can withstand a more harsh environment, such as the acid soils and aluminum toxicity of the lateritic soils of the Cerrado of Brazil. An intensive breeding and selection program is yielding varieties suited to this type of soil. The Cerrado includes 200 million hectares of land with adequate rainfall and tillable soil. In a decade the Cerrado will probably be producing significant quantities of grain. In addition, success in the Cerrado might speed development of other areas of the tropics where similar conditions prevail.

The maize program at CIMMYT is receiving an effort comparable to that devoted to wheat and triticale. Maize can be grown under a broader set of environmental conditions than wheat and it is particularly adapted to the tropics. Wild types have been growing in many parts of this hemisphere. More than ten thousand lines have been placed in maize banks. An intensive breeding program is creating higher yielding varieties. Norman Borlaug, the Nobel laureate who has specialized in developing strains of wheat, believes that in five to ten years production of maize will outstrip wheat production worldwide. Maize has the additional advantage of yielding about 50 percent more grain per hectare than wheat in many localities. A factor that may speed the adoption of maize is the development of a more nutritious variety. About a decade ago, the availability of a high-lysine, high-tryptophan line was announced. However, the plants then available were deficient in yield and disease resistance. Patient work over a decade at CIMMYT has led to varieties that are nutritionally balanced and also high yielding and disease-resistant.

In addition to maintaining the basis for a response to threats of crop failure due to disease and to improving prospects for additional food, CIMMYT maintains a training program for agriculturists from many countries. Total annual cost of all these efforts is $20 million, only a fraction of which comes from the U.S. government. Seldom has so modest a sum produced such significant results.

---

5 August 1983

Much of the tropics is favored with tillable soils, abundant sunshine, and adequate rainfall, but vast areas have been little used. This is particularly true of Amazonia, which constitutes the world's greatest frontier area. However, an increasing tempo of exploitation of its rain forests is now making management of the region a controversial issue. Some ecologists have taken the position that the wilderness should be left intact, and they can cite mistakes that have been made. Thus far, the damage is relatively limited when account is taken of the vast area of the Amazon Basin. But it is clear that large-scale attempts at exploitation will occur and that unless these are based on scientific knowledge, an inferior outcome will ensue.

Much of Amazonia is covered by luxuriant forests. But for the most part the trees cover a very poor soil. Over millions of years, warm, naturally acid rain has leached nutrients from the soil, leaving such barren materials as kaolinite, sand, and aluminum and ferric oxides. The pH of the soil is low, cation exchange capacity is small, and toxic ionic aluminum is usually present. In a typical rain forest, 70 to 90 percent of the phosphorus, potassium, calcium, and magnesium present in the ecosystem is located in living phytomass. As leaves or other materials fall to the ground, they are soon mineralized, and the product is quickly taken up by roots. Areas of the rain forests have been exploited for crop growth, pasture, and logging.

Subsistence farmers make a poor living from slash-and-burn agriculture. This involves cutting down a patch of the forest and burning the wood in situ. The ash contains most of the nutrient minerals, and its alkalinity raises the pH of the soil. A crop of upland rice can be obtained. However, many of the nutrients are lost by leaching, and after a year the patch is usually abandoned. The succeeding vegetation is inferior. Typically, twenty years elapse before the slash-and-burn cycle is repeated. That kind of practice depletes the soil and cannot support a prosperous economy. In contrast, immediately following slash and burn, the land can be successfully used for cropping or pasture provided good management and fertilizers are available. In a region of tropical rain forests of the Amazon Basin in eastern Peru, fertilization permitted a continuous three-crop-a-year agriculture. Rotation of crops held down damage from pests, and yields were excellent. Soils were monitored, and after seven years a marketed improvement was noted.

Experience in Brazil has shown some of the consequences of converting forests into pastures. If appropriate grass seed is sown in the year following slash and burn, a pasture can be established. However, unless fertilizer (particularly phosphate) is applied, the quality of the pasture deteriorates and substantial areas have been abandoned.

Of a total of about seventeen hundred species of trees found in Amazon forests, only eight have a commercial market. This makes harvesting costly and destructive. At the same time, the natural forests are senile. They produce only 1 to 5 cubic meters of biomass per hectare per year. In contrast, managed forests could produce 30 to 80 cubic meters per hectare per year.

For the use of fertilizers to be practical, it is necessary to have cash crops, markets, and transportation. Successful large-scale development is most likely to occur in the limited areas where soils are relatively fertile and accessible to transportation, as on the floodplains of rivers. An important constraint on satisfactory development is lack of enough expert scientists versed in the special problems of tropical soils and agriculture. At a recent symposium on Amazonia in Belém, Brazil, it was clear that there are first-class scientists in the region. However, there are large gaps in knowledge about crops that require minimal fertilization, and the governments are decreasing support for research. To achieve some measure of the productive potential of the region will require a much more intensive scientific effort coupled with many agricultural and silvicultural demonstration farms and corresponding extension services.

---

9 May 1975

When food is abundant, it is wasted or treated as a commodity. But when food is scarce, it is regarded as the staff of life and its distribution becomes a highly emotional issue. Food production worldwide is increasing faster than population, but distribution is uneven, reserves are limited, and bad weather conditions could lead to widespread famines. Prospects of poor crops universally are not great but variations in harvests good and bad will continue to place us on an emotional roller coaster. The United States and much of the rest of the world seem to be entering a phase in which supplies of food will be adequate. It is a good moment for stocktaking.

The circumstances surrounding world production and distribution of food involve complex interactions of politics, economics, weather, and other factors. When we speak of the developing countries and lump together their statistics, we create a composite that hides more than it reveals. Some are wealthy, some are poor, some are food exporters, some are highly vulnerable to vagaries of weather. Two examples, India and China, point up differences in food production and distribution in two nations whose combined population is about half that of the developing countries. China, possessing considerably less agricultural land than the United States, feeds a population about four times as great. Key factors are irrigation and transplanting techniques that permit as many as twelve crops of vegetables to be grown annually on some plots of land.

An extremely important aspect of food is nutrition. This is especially true for vulnerable groups such as infants and young children. Prudence dictates that in the future we determine how best to cope with bad weather and other adverse conditions that have often caused famines. Ultimately, exponential growth of population must diminish and cease. Cessation of growth may come through calamities such as pestilence or nuclear war, it may come through starvation, or it may come through a gradual

Food and Nutrition

change in attitudes. If the latter road is to be followed time is required, for customs usually evolve slowly.

Much could be accomplished by more effective use of what is already known. Crop yields in some countries are smaller than in corresponding regions where better practice prevails. Losses due to pests before and after harvest are substantial and some of these are avoidable. Large areas of the tropics are not being cropped because their soils are not now suitable. Better understanding of how to deal with such problems is on its way. One means of increasing yields as well as assuring crops is irrigation. A new system involving underground application of water has substantial potential.

Currently only about 1 percent of the solar energy falling on an area is fixed by photosynthesis. Basic biological research may lead to better efficiencies. This may come about through genetics. One path is the creation of new species of plants by transfer of DNA. In a time of shortages of energy, improvement of nitrogen fixation by plants is also desirable. Indeed, the effort to make more effective use of solar energy has relevance that goes beyond food, for ultimately the world must come to depend largely on renewable resources to fill its many needs.

4 April 1969

Children reared in poverty tend to do poorly on tests on intelligence. In part this is due to psychological and cultural factors. To an important extent it is a result of malnutrition early in childhood. This matter is discussed in a recent article in *Science* and in a report of an International Conference on Malnutrition, Learning, and Behavior [N. S. Scrimshaw and E. Gordon, eds., *Malnutrition, Learning, and Behavior* (MIT Press, 1968)]. Both publications survey the effects of deprivation at an early age. It seems likely that millions of young children in developing countries are experiencing some degree of retardation in learning because of inadequate nutrition, and that this phenomenon may also occur in the United States.

Because of complex social and psychological factors associated with malnutrition, it is not easy to assess the effects of dietary deficiencies in man. However, observations in underdeveloped countries, coupled with studies on animals, provide substantial evidence. In rats and pigs the brain reaches 80 percent of adult size by normal weaning time. At that stage, body weight is 20 percent of that at maturity. During the period of rapid growth the brain is vulnerable to nutritional damage. A relatively short period of undernutrition results in smaller brain size at maturity even if the animals are maintained on a good diet after weaning. Changes in brain size are accompanied by persistent anatomical and biochemical changes.

In humans, the brain of the infant attains 80 percent of adult weight by age three, when the body weight is about 20 percent of that at maturity. Thus the animal experiments suggest that good nutrition during the first three years of life is particularly important.

In some countries undernutrition involves a deficit in calories, proteins, and vitamins. Usually lack of protein is the most serious problem, but vitamin deficiencies are also important. Throughout much of the world, grains are the principal sources of protein. These do not contain a full complement of all the essential amino acids—lysine is usually in short supply. Many individuals, even in the United States, who have adequate caloric intake primarily from low-cost foods may be malnourished. This possibility has become a matter of increasing concern to the federal government.

Desirable objectives for a U.S. food program have been described by Dr. Aaron M. Altschul of the Department of Agriculture: no one must go hungry; hunger or malnutrition must not be a deterrent to economic development; and the American diet must provide for optimum health throughout life.

These are laudable goals, but they will not be easily attained. Enough food can be provided, but that is not the whole answer. The consumer must choose to eat nourishing foods. In part, the problem can be met by enriching commercial food products. For example, cereal foods such as wheat flour or products made from it are markedly improved by the addition of lysine (0.2 percent) and of needed vitamins and minerals.

Good nutrition is desirable at all ages, but it is evident that the first three years are crucial. To respond to this need, the federal government has begun to distribute, in some places, a special supplementary food package to new and expectant mothers, through health centers and clinics. This program should be expanded. However, unless mothers understand how to care for their young, bountiful supplies of food will not suffice. There is real need for general education in the basic principles of nutrition and for continuing effort to assure that no one, and especially no infant, fails to develop properly because of malnutrition.

---

5 February 1971

Our nation has been blessed with high-quality, dependable supplies of low-cost food and fiber. But few people are aware of the never-ending battle that makes this possible. Among the insects, about ten thousand species are known to damage crops, forests, or livestock. If left undisturbed, some of these pests would take more of the crop than would the farmer. Agriculture must cope not only with native enemies but also with invaders from abroad. Increasing air travel has expanded the probability of bringing in pests from all over the world. For these, natural controls such as parasites are often absent.

During more than two decades, agriculture has depended heavily on chemical pesticides such as DDT. But even in the 1950s it was apparent that DDT was not the final answer; resistant insect forms were emerging. The use of degradable chemicals such as the organic phosphates and carbamates has been expanding. However, these are acutely toxic to warm-blooded animals, including man, and also to useful insects such as bees. For the long term it is desirable to develop methods of control that minimize the use of broadly toxic chemicals.

Some of these methods antedate the use of the chemical pesticides—for instance, the use of insect-resistant varieties of plants and the fostering of natural enemies. Promising methods of control that have recently been developed include the use of sex attractants and the release of competitive radiation-sterilized males. About twenty of the sex attractants have now been identified. Of particular importance is that of the gypsy moth, an insect that has been ravaging a rapidly increasing fraction of the hardwood trees of northeastern United States. The synthesis of the attractant and its use in conjunction with localized applications of insecticides give promise of timely intervention in what otherwise might be a dreadful destruction of vast stands of trees.

At the AAAS meeting in Chicago, E. F. Knipling from the Department of Agriculture described proposals to exploit techniques that he had previously used successfully to eradicate the screwworm in Florida. These same techniques have been employed to minimize the population of screwworms in Texas and adjoining states. In that area, eradication is impossible because of reinvasion from Mexico. However, the pest is more than 99 percent controlled at an annual cost of about $6 million and with annual benefits estimated at $100 million. Knipling pointed out that a limited number of pests account for a very major fraction of all crop damage. Under especially close study are the boll weevil, the codling moth, and the corn borer. The boll weevil alone causes losses of about $200 million a year, and about $70 million is spent by growers for chemical control. About one-third of all insecticides used for agricultural purposes is employed to control this single insect. Knipling discussed plans for a large-scale demonstration experiment to suppress populations of this insect in an area 250 kilometers in diameter. The plans involve a combination of treatments to diminish the overwintering population and an intensive program in the spring to trap insects by use of sex attractants. The survivors would be subjected to repeated release of one hundred times their number of sterilized males.

The suppression or elimination of the use of DDT and related pesticides will constitute a hardship to many farmers. Prospects are reasonably bright, however, that satisfactory methods of specific pest management will be devised, and these methods will be more in keeping with our desire to minimize man's destructive impact on the environment.

18 December 1964

There is worldwide interest in means for obtaining more fresh water. President Johnson recently indicated his views, in part, by quoting a statement made by President Kennedy: "There is no scientific breakthrough, including the trip to the moon, that will mean more to the country which first is able to bring fresh water from salt water at a competitive rate. . . ."

Much of the U.S. effort aimed at desalination of water has been conducted under the Office of Saline Water of the Department of the Interior. The OSW has sponsored some imaginative work—for example, development of a reverse osmosis process. Given a membrane permeable only by water, it is possible to obtain fresh water from sea water by exerting a differential pressure of about 24 atmospheres on salt water. In turn, it is theoretically possible to obtain a cubic meter of fresh water from sea water by the expenditure of about 0.7 kilowatt-hours of energy, the minimum amount for a reversible process. The present cellulose acetate membranes, however, are not perfect. A pressure of 100 atmospheres is required to produce water at the rate of 370 liters per square meter of membrane surface per day. The effluent, while potable, is not entirely free from salt. The membranes have only a few weeks of service life.

Another process involves formation of clathrate compounds. If propane, under pressure of 3 atmospheres, is bubbled into sea water at about 6.5°C above the freezing point, a light solid composed of 17 molecules of water to one molecule of propane is formed. Salt is excluded. The solid can be removed and the water readily obtained. In a practical process, energy consumption might amount to between 3.5 and 12 kilowatt-hours per cubic meter of water.

Another method, which may be used practically, is freezing. It is attractive because the latent heat is small, and low-temperature processes are relatively free from corrosion problems.

Desalination of Water

Energy cost for a practical freezing scheme has been estimated at 12 kw-hr/m^3.

Without heat exchange, the energy requirement for distilled water is about 700 kw-hr/m^3. With heat exchange, this figure has been reduced to 70 kw-hr/m^3 in "demonstration plants," and it could be reduced to somewhat lower values through greater investment for equipment. The price of obtaining fresh water from sea water is heavily dependent on the cost of energy. Recently it has been pointed out that very large nuclear reactors could be particularly efficient. In situations where a dual-purpose electricity and water plant are feasible, costs for distilled water have been estimated at about 6 cents per cubic meter. (Costs of raw fresh water in coastal regions of the United States range from about 0.1 to about 6.0 cents per cubic meter.) In high-cost areas such as southern California, some of the needs probably could be met by desalination through distillation. In the eastern half of the nation, where total supplies are adequate and cheap, better utilization of natural supplies is the practical solution.

It is to be hoped that a balanced approach will be made to the water problem. Adequate emphasis should be given to research and to development, particularly of promising new processes. At the same time it should be remembered that most of this nation's water problems will be solved by wise use of what is naturally available.

1 October 1976

Space research is yielding an increasing body of new knowledge and practical applications. Conspicuous examples are the results from the Vikings and the success of communications satellites. Yet an important multidisciplinary, multinational series of global studies is receiving little notice. The enterprise was born 23 July 1972 with the launching of the first Earth Resources Technology Satellite (ERTS-1) by the National Aeronautics and Space Administration. Since that time, most of the solid earth has been imaged, much of it repeatedly, and scientists and engineers have been exploiting the data. Some of their findings are displayed in a recently published professional paper of the U.S. Geological Survey [R. S. Williams, Jr. and W. D. Carter, eds., *ERTS-1: A New Window on Our Planet*] that contains about ninety articles on eight major topics, including applications to cartography, geology and geophysics, water resources, land-use mapping and planning, environmental monitoring, conservation, and oceanography.

Utilization of data from ERTS-1 has been widespread. Some one hundred nations are participating. Application to cartography and geology and geophysics are numerous. This was evident at the International Geological Congress in Sydney, Australia, during August 1976.

ERTS-1 is in a near-polar orbit about 918 kilometers above the earth. It circles the earth every 103 minutes and views each area of the earth every eighteen days. ERTS-1 obtains images in four bands of the spectrum: 0.5 to 0.6, 0.6 to 0.7, 0.7 to 0.8, and 0.8 to 1.1 micrometers. Data from the images can be combined to produce false-color pictures or can be analyzed by computer.

A particular advantage of ERTS-1 is the periodic coverage. Images obtained at different times can be compared and short-term or long-term changes evaluated. For example, variations in the chlorophyll content of fields can be detected. Healthy plants contain more chlorophyll than sickly ones. Thus, information on crop prospects can be garnered. The images also show the extent of forests. By comparing winter and summer data, the abundance of deciduous trees can be established. Yet another example is in management of water resources. In many areas of the world, regulation of reservoirs fed by melting snow is vital. By studying successive ERTS images and related ground data, better management is possible.

Longer-term periodic coverage may prove to be particularly important. One of the chapters of the USGS professional paper is devoted to the mapping of Antarctica, especially its coastal areas. The images will be part of a historical record that will show changes in the size, shape, and position of such features as ice shelves, glaciers, and ice tongues. With continuing monitoring of such features, ERTS-1 or its successors may well provide the first substantive indication of global climatic changes.

Another important advantage of the ERTS system stems from the height at which the satellite moves. Thus, an image covers an area of 34,000 km^2, about one thousand times that covered in an aerial photograph from a high-flying plane. In consequence, large-scale features of the earth have been identified that had previously gone unnoticed. Of particular importance to economic geology is the discovery of large-scale linear and curvilinear features.

Scanning the articles in the USGS publication, one can find evidence that the ERTS-1 venture is more than simple exercise in photography whose potential will be quickly exhausted. It is clear that in four years scientists and engineers have found many other uses for the data. Moreover, with experience the power of the applications has increased. For example, images obtained from rocky terrain tend to appear as only slightly varied shades of gray. However, with electronic data processing, it becomes possible to identify the delicate shades as different rock formations.

The scientists using the ERTS-1 images are enthusiastic. Many feel that such satellites will be of great help to all nations in coping with natural resource and environmental problems, and that the ERTS data could not have become available at a more propitious time.

12 June 1970

Early this spring, Joseph W. Spelman, chief medical examiner of the city of Philadelphia, addressed medical colleagues on the topic of sudden death from heroin. To a shocked audience he showed photograph after photograph of victims with needles remaining in their veins who had died after self-administration of drugs.

In New York City, among the estimated one hundred thousand heroin addicts, more than nine hundred fatalities due to drugs occurred in 1969. In that city, for the age group fifteen to thirty-five, drug abuse is now the leading cause of death. According to Michael M. Baden, deputy chief medical examiner, the majority of fatalities are due to an acute reaction to the intravenous injection of a mixture containing heroin. The mechanisms involved in the deaths are not clearly established: overdosage has been suggested by some investigators; others speak of an allergic reaction. A survey of practices attending the production, distribution, and usage of heroin leaves one amazed that the death rate is not higher. The method of illicitly extracting morphine from opium is crude. The impure morphine is subsequently acetylated to heroin in secret laboratories, mainly in France. Purity of product is of the order of 90 percent. Subsequently, the heroin passes through a complex distribution system and is adulterated repeatedly in unsterile conditions with a variety of additives, including quinine, mannitol, and other white powders.

The Office of the Chief Medical Examiner of New York City analyzed 132 street samples of drugs, all of which supposedly contained heroin. They found that 12 samples contained no heroin at all, and among the remainder the concentration of the drug ranged from less than 1 to 77 percent. Variability in the amount of the drug could be responsible for many fatalities. A user accustomed to a low concentration is likely to die from an injection of almost pure heroin.

Hard-core addicts subject themselves to more than one thousand intravenous injections each year, and they are thus

Death from Heroin

exposed repeatedly to possible antigens in the crude heroin or in its adulterants. In addition, the repeated use of unsterile drugs, unsterile equipment, and unsterile technique leads inevitably to human wreckage. In a description of the major medical complications of heroin addiction, Donald B. Louria and his colleagues have identified the most common medical problem as liver damage arising from hepatitis. Other organs that are particularly subject to attack include the heart and lungs. Infection of the heart, though not so frequent as hepatitis, is more often fatal.

Drug abuse, which was once predominantly a disease of Harlem, is now a plague that is spreading to the suburbs. Drug use has been glamorized, while descriptions of the dreadful consequences have been muted. Parents and educators must inform the young of the corpses and of the physical wreckage. Despite warnings, adventurous youth will sample the illicit—and many will be hooked. The number of addicts is already estimated at two hundred thousand, and the annual cost of their drugs at $5 billion. With so much at stake in lives and in money, the nation should increase its efforts to curtail drug abuse and to find better ways to rehabilitate addicts. Two relatively new methods seem promising. One is the use of methadone. A second approach is a psychiatric one, which emphasizes attitudinal changes and utilizes ex-addicts to give emotional support to addicts who wish to stop. Determined and imaginative effort might well disclose even better methods. This nation should provide the necessary funds to move vigorously against a spreading plague.

7 November 1969

By accident or by design, humans are being exposed to an increasing number of chemicals. Each year thousands of new compounds are synthesized, and people subsequently are subjected to them in many ways through agents that include air, water, food, drugs, and cosmetics. Acute toxicity of pure substances can be determined quite readily; synergistic effects are more difficult to evaluate. However, the major causes for concern are delayed effects such as cancer or genetic mutations. Because of the long times necessary for manifestation of damage, the hazards attending long-term exposure to most chemicals are not known.

The primitive level of our knowledge was demonstrated recently in events attending the banning of the use of cyclamate. This artificial sweetening agent has been in use for about two decades. Short-term and long-term toxicity of the chemical had been tested repeatedly, but only now have studies of eighteen-month exposures at high dosages been conducted; such exposure resulted in cancers of the bladder in rats. There is no evidence that cyclamates have caused cancer in man, but the full effects of the mass exposure cannot be known for twenty years or more.

Cyclamate is only one of hundreds of chemicals—some synthetic, others natural products—that have been ingested by man. Often, for what seem rather trivial benefits, risks of undetermined magnitude have been incurred by substantial fractions of the total population. Thus far we have escaped disaster. Indeed, insofar as they can be interpreted, the long-term trends in the incidence of cancer are to some extent reassuring. Although there has been an overall increase, the major factor in the increase has been lung cancer associated with smoking. On the other hand, incidence of cancer in the stomach and liver has decreased sharply.

The public rightly expects the government to take measures necessary to insure its safety from environmental hazards, and legal authority exists for doing part of the job. No new food additive can be introduced in interstate commerce unless the manufacturer produces evidence that it is safe. The applicable phraseology includes the following statement from the so-called Delaney Amendment: "no [food] additive shall be deemed to be safe if it is found to induce cancer when ingested by man or animal. . . ."

The effect of the law has been to subject new additives to close scrutiny while permitting continued use of substances that were being used prior to 1958, even if they had never been thoroughly tested. Moreover, many foods are known to contain natural components which, when given in massive doses, have been found to produce cancer. In addition, smoked foods usually contain highly carcinogenic compounds.

The government utilizes unevenly its power to protect the public. If the nation were truly concerned with cutting the incidence of cancer, it would mount a massive campaign against cigarette smoking and it would closely examine all substances that are ingested, including even those that have long been used.

Knowledge concerning the factors leading to cancer is fragmentary and hard to acquire. However, we have not been as vigorous as we might be in obtaining it. Long-term animal research has not enjoyed sufficient support. Epidemiological studies in man should be expanded. Around the world, widely differing incidences of cancer are noted. Careful analysis of this phenomenon should be made. In addition, more comprehensive studies should be initiated in the United States, including a follow-up on individuals exposed to large doses of cyclamate.

23 September 1983

Bruce Ames has written about the increasing body of evidence that large numbers of potent carcinogens arise from natural processes. Mutagens are present in substantial quantities in fruits and vegetables. Carcinogens are formed in cooking as a result of reactions involving proteins or fats. Dietary practices may be an important determinant of current cancer risks.

Ames describes the role of plant materials as follows: Plants in nature synthesize toxic chemicals in large amounts, apparently as a primary defense against the hordes of bacterial, fungal, and insect and other animal predators. Plants in the human diet are no exception. The variety of toxic chemicals is so great that organic chemicals have been characterizing them for over one hundred years, and new plant chemicals are still being discovered. Recent widespread use of short-term tests for detecting mutagens, and the increased testing of plant substances for carcinogenicity in animals have contributed to the identification of many natural mutagens, teratogens, and carcinogens in the human diet.

Safrole and related compounds are present in many edible plants. Safrole is a carcinogen in rodents and some of its metabolites are mutagens. Oil of sassafras, once used to flavor some root beer, is about 75 percent safrole. Black pepper contains about 10 percent by weight of a closely related compound, piperine. Extracts of black pepper at a dose equivalent to 4 milligrams of dried pepper per day cause tumors at many sites in mice. Many hydrazines are carcinogens and mutagens, and large amounts of them are found in edible mushrooms. One carcinogenic hydrazine is present in the false morel at a concentration of 50 milligrams per 100 grams. It causes lung tumors in mice at a level of 20 micrograms per mouse per day.

Carcinogens and mutagens are present in mold-contaminated foods such as corn, nuts, peanut butter, bread, cheese, and fruit. Some of these contaminants, such as aflatoxin, are among the most potent known carcinogens and mutagens. Nitrosamines and nitroso compounds are suspect as causative agents of stomach and esophageal cancer in humans. In the digestive system these nitrogen compounds are formed from nitrate and nitrite. Beets, celery, lettuce, spinach, radishes, and rhubarb all contain about 200 milligrams of nitrate per 100-gram portion.

Rancid fats are possible causative agents of colon and breast cancer in humans. These forms account for a substantial fraction of all the cancer deaths in the United States. Unsaturated fats are easily oxidized on standing and in cooking to form mutagens, promoters, and carcinogens. Among the numerous products of such oxidations are fatty acid hydroperoxides and cholesterol epoxide. Thus the colon and digestive tract are exposed to many fat-derived carcinogens. Human breast fluid can contain high levels of cholesterol epoxide.

Burnt and browned materials formed by heating proteins during cooking are highly mutagenic. Chemicals isolated from such products have been found to be carcinogenic when fed to rodents. In addition, the browning reaction products from caramelization of sugars, or the reaction of amino acids and sugars during cooking, contain a large variety of DNA-damaging agents.

The view that dietary practices might be a causative factor in cancer is not new. Epidemiologists have noted marked differences in cancer rates between population groups. Effects from changes in diet following migration have also been observed. Results of current studies are beginning to delineate more sharply specific causative agents. When more definitive information is available, it should be possible for prudent people to choose fruits and vegetables that present minimal hazards. In the meantime, there is persuasive evidence that charred meats and rancid fats should not be part of the diet.

5 October 1979

When people are unable to evaluate the magnitude of risks in a situation, many are inclined to fear the worst. This has been especially true in attitudes toward the health hazards associated with chemicals. A tendency for some people to be "spooked" has been exploited by opportunists who have talked of a cancer epidemic allegedly created by chemicals. Some chemicals in the workplace are carcinogenic, but their contribution to cancer prevalence is small in comparison to that of smoking and probably to that of natural substances in raw and cooked foods.

Talk of a cancer epidemic seems to owe much to misinterpretation of a statement made by John Higginson of the International Agency for Research on Cancer. He has written, "between 80 and 90 percent of all cancers are dependent directly or indirectly on environmental factors." However, Higginson recently emphasized that the principal basis for his estimate was such factors as lung cancer from smoking and carcinogens such as aflatoxins in foods. He estimated that 1 to 5 percent of cancers are due to the occupational environment.

The cancer stampede has created opportunities for people in politics and others yearning for publicity; it has also been tempting to some scientists in government. A group from the National Cancer Institute, the National Institute of Environmental Health Sciences, and the National Institute for Occupational Safety and Health gave then Health Education and Welfare Secretary Joseph Califano alarmist estimates that he used in a speech on 15 September 1978. The speech, which was widely quoted, included the statement that during the next thirty to thirty-five years, 17 percent of all cancer deaths will be associated with previous exposures to asbestos. The backup report for the estimate (not in a peer review journal) was flimsy. For example, an important component of the cohort cited as being at future risk was people who had worked in shipyards during World War II. The authors provided no data about the age distribution of the workers. However, because of the military

Cancer: Opportunism and Opportunity

draft, the average age of working males was probably forty to forty-five. By now, these people must already have made most of their contribution to cancer statistics. The report was attacked by many epidemiologists. Richard Doll, Regius Professor of Medicine at Oxford, wrote, "I regard it as scientific nonsense." An editorial in the British medical journal *Lancet* criticized the report, concluding with, "it is sad to see such a fragile report under such distinguished names."

The effort to prove a big role for industrial chemicals diverts attention from what is probably the best hope for reducing cancer incidence—careful study of foods and effects of cooking. It has long been known that the incidence of gastrointestinal cancers is highly variable, both temporally and geographically. For example, a variation of the incidence of esophageal cancer by a factor of one hundred has been observed in China, and there is a strong correlation between prevalence and food. Yet expenditures to examine the role of food in cancer have been comparatively small here. Leadership has been seized by Takashi Sugimura, director of the National Cancer Center in Tokyo. He has used the Ames test and other related procedures to detect mutagens and carcinogens in raw foods. He has also conducted experiments on the effects of elevated temperatures on proteins and amino acids and has observed the resulting formation of strong mutagens and carcinogens. He has pointed to differences in food preparation as the reason why incidence of stomach cancer in Japan is twice that here. Sugimura and colleagues have also studied mutagenic effects of flavonoids, including the highly mutagenic quercetin and kaempferol, which are present in many edible plants, vegetables, and fruits.

All people ingest daily the mutagens and carcinogens of food. Far fewer are exposed to hazards in the workplace. It is now feasible to monitor the mutagens naturally present in food and to evaluate changes arising from cooking. A comprehensive investigation of foods and the products of cooking should have high priority.

---

14 May 1976

Private and government expenditures for health are now devouring a substantial part of the gross national product. In spite of considerable evidence that the health care system is not cost-effective, the prospect is for continued growth. In 1950 funds spent for health were $12 billion. In 1975 they reached $118.5 billion. Until 1965 the government spent only nominal amounts for health care. With the advent of Medicare and Medicaid, the federal treasury became an engine of inflation of health costs. Federal expenditures rose rapidly to $34 billion in 1975, and this led to enhanced costs for the private system also.

Articles have appeared in which authors seek to identify specific causes for the increase—for example, soaring hospital costs. For the most part they miss the main point. When the major fraction of medical costs is borne by a third party, demand for care is practically infinite. Patients urgently seek treatment, even surgery, on the basis of trivial symptoms. Distraught relatives hope to prolong the lives of moribund loved ones. Fearing malpractice suits, many physicians practice medicine defensively, ordering more tests and procedures than they otherwise would.

The public assumes that large expenditures for health care will bring better health. This assumption is questionable. During the early part of this century life expectancy in the United States steadily increased, but it reached a plateau in 1954. In 1967 W. H. Forbes explored the relation between national expenditures in behalf of health and actual results. He concluded that we could halve or double the total expenditures without changing longevity. This was in a year when only $42 billion was spent.

Since 1967 others have pointed out that most of the deaths in the age range ten to seventy either are due to degenerative diseases or are fatalities arising from accidents, suicide, or homicide. The big killers are coronary heart disease, cancer, and stroke. Treatment of these diseases is often costly. Their

Cost-Effective Health Care

incidence is related in part to life-style—for example, sedentary living, poor diet, obesity, smoking.

Because treatment of degenerative diseases is not uniformly successful and since the course of some of them can be altered by changes in the patient's behavior, there is increasing interest in preventive medicine. Frederick C. Swartz, M.D., has stated that "our greatest health problem is in the physical fitness of the Nation. Here the answer is the simplest and the cheapest, has the greatest application, and its reflection on the reduction of morbidity and mortality rates would be immediate and tremendous. It is entirely possible that a well-practiced physical fitness program begun early in life would increase life expectancy by ten years. . . ."

Studies seem to show that longevity depends on a combination of factors. Prominent among them are good nutrition, weight control, abstention from excessive drinking or alcohol and from cigarettes, and getting enough exercise and sleep. Faced with the prospect of giving up smoking and engaging in vigorous exercise, many people would just as soon take their chances. However, others would like to pursue a more prudent course. They would be encouraged to do so if they had specific information about the effort required to increase their life span.

Substantially better health cannot be bought with $118.5 billion. Isn't it time the nation began to pay more attention to approaches that promise great improvement at little cost?

17 May 1963

The Soviet ability to launch large missiles has been misinterpreted as indicating superiority over the United States in scientific matters. Actually there are few areas of science in which the Russians excel. Despite their ability to place large pieces of hardware in orbit, their contributions to space research have been meager. They have nothing to match our Mariner II mission results, and their exploration of regions closer to the earth has been less intensive than ours. In high-energy nuclear physics our discoveries are unmatched, as is our progress in maser-laser studies and in semiconductors. In most areas of chemistry the Russians are behind us; plastics and petrochemicals are outstanding examples. In the exploitation of radioactive isotopes much of the Russian work is mere repetition of our research. In biochemistry, biophysics, and molecular biology we are superior. The Russians have achieved nothing like our progress in deciphering the genetic code or in determining amino acid sequences in proteins.

Innumerable examples could be given; we compete on countless frontiers of science that, in sum, are vastly more important than space. On many of these frontiers, such as solid-state physics, advances are crucial to future economic and military strength. Fundamental research is now often quickly followed by practical applications. A substantial fraction of today's commerce is based on discoveries of the last two decades. It is of interest to compare American and Russian competition in world markets in items involving science and technology. By this yardstick the Soviet Union is a third-class power. It is no match for Western Europe, the United Kingdom, or Japan.

The Japanese are competing in technological areas requiring first-class scientific competence. Their electronics products such as transistor radios and television sets are selling for less than ours on our own soil. To a degree this reflects cheaper labor, but only in part. The production of transistors and other solid-state electronic components involves sophisticated tech-

International Competition in Science

nology. Even the cheapest labor is no substitute for scientific ability in this field.

Western Europe is far stronger scientifically and technologically than the U.S.S.R., and the Western Europeans are rapidly closing in on us. If present trends continue, it will be only a matter of a few years before they achieve supremacy.

Western Europeans have long proved that they are, individually at least, as competent scientifically as we. They have made a remarkable recovery from the effects of World War II and are again in a position to challenge us. In the contest they have two advantages. Research costs them about one-fourth what it costs us, and proportionately less of their talent is occupied with military and space efforts. Leaders of industrial research in this country are increasingly concerned with the overpowering competition of government-financed programs for first-class scientific talent. One research director told me recently, "We need good people, but my company can't compete with projects paid for by the U.S. treasury."

We have chosen to stake our national prestige in a propaganda contest with the Russians in one of the few major areas of technology where they have an edge over us. In the meantime we fail to note that the Western Europeans are getting ready to walk away with the trophies that really count.

20 October 1967

Science and technology have given man vastly enhanced powers to achieve socially desirable purposes, but the exercise of some options opens the way to dangerous abuses. A particularly worrisome development is the increasing use of science and technology in the invasion of privacy. The magnitude of the present problem and its future scope have been surveyed in a thoughtful volume entitled *Privacy and Freedom* by Alan F. Westin, professor of public law and government at Columbia University.

The author begins by discussing the individual's basic need for a minimum of privacy. He links the human requirement to the well-known territorial behavior of many animals. He acknowledges the need for society to exert controls on individuals and groups. However, he points out that occasional privacy is essential to the health of individuals as well as to that of most organizations.

Many means for invading privacy are now available. These include new devices and techniques for physical surveillance, psychological surveillance, and data surveillance. These developments provide sophisticated help for the merely curious. They furnish technological assistance to modern Peeping Toms. They can make life easier for a growing army of private detectives. The devices find wide application in business, both for legitimate surveillance and for industrial espionage. They are employed by many branches of government, sometimes in questionable ways. They provide governments with tools that one day may be used to bring about George Orwell's 1984.

Some of the techniques are old and familiar. They include wiretapping and use of hidden microphones and cameras. These methods have been improved, however, as technology has advanced. Thus, solid-state electronics permits miniaturization. Microphones the size of a sugar cube ($10) or the size of a pea ($100) are available.

Photographic techniques also have been improved. Special screens can be installed in walls, which seem opaque to persons inside a room, but permit an observer outside the room to see and photograph events within it. A more advanced method is to employ infrared light in conjunction with appropriate panels and cameras. This permits photography in what appears to be total darkness. A possible future development is the use of computers to analyze telephone conversations. In principle, all such conversations could be monitored and recorded. Computers might be used to identify characteristics of a person's voice. Then the computer could search out all phone conversations in which that individual took part, even when the call originated from a pay station. In a day when so much of society's business is transacted by phone, a powerful tool for the invasion of privacy would be available.

The computer is a key to a related kind of invasion of privacy, in that it quickly makes available detailed information about a large number of individuals. All of us have filled out dozens of forms. What is new is the ability of a centralized bank of computers to store the massive amounts of data and quickly retrieve them.

The ultimate extent and consequences of erosion of privacy can be only dimly perceived. However, it is clearly desirable to seek technical and legal means of curtailing the use of what might otherwise become instruments for the destruction of our freedom.

7 November 1980

Profound changes are occurring in the creation, processing, storage, retrieval, and transmission of scientific numerical data. The power and convenience of electronic instrumentation has led to its wide use in laboratories. Digitized data from such equipment can readily be stored and processed. They can be retrieved and transmitted to other equipment both here and abroad through telecommunication networks. In many areas of science and technology, large electronic data bases are being created that can also be tapped through the networks. Improvements in very large scale integrated circuits and memory devices, together with development of additional software, guarantee a great expansion of the role of electronic data.

New electronic devices have made possible experiments and observations not previously attainable, and also the accumulation of data at unprecedented rates. This is true throughout the natural sciences. The exploration of Jupiter by the *Voyager* spacecraft was completely dependent on electronic sensing device, communication of signals to the earth, storage of the data in memories, and subsequent machine processing. The Geosynchronous Environmental Operational Satellite measures visible and infrared spectra of the earth's disk every 30 minutes and produces 2×10^{11} bits of data every day.

Many processes in nature occur in very short times. An important research frontier today is picosecond chemistry. Through the use of lasers and electronic sensing devices, much information is now being gathered about excited states of atoms and molecules. Details of the mechanisms of photosynthesis are being studied. When light falls on a plant, excited states are produced, electrons are transferred, spectral changes take place. These phenomena occur in time spans of microseconds or less.

New instrumentation has had profound effects on analytical chemistry. The most striking one has been to create the capability of identifying and measuring very tiny amounts of substances. By employing a combination of gas-liquid

Electronic Data Bases

chromatography and mass spectrometry, biochemists have been able to isolate and measure 0.1 picogram of a hormone. With other equipment, analyses can be made much more rapidly than heretofore. A new spectrophotometer produces an entire spectrum from 200 to 800 micrometers in only 1 second.

A major hazard in hospitals is errors of transcription, which sometimes run as high as 5 percent or more. Modern hospitals try to avoid such errors in clinical laboratories by using electronic devices and storing results in a computer. A similar situation exists in pharmaceutical laboratories, which must maintain records of exemplary quality. Every measurement possible, such as weighing, is conducted with equipment that ties into the computer.

Electronic storage of digital data is the only feasible means of dealing with information in areas of science where it is produced at such a great rate that placing it on paper would be impractical. In addition, once the massive amounts of data are in machine-retrievable form, they can be processed and analyzed quickly, and with a thoroughness beyond human capability.

Computers can communicate with each other, and this is being facilitated by public and private networks. Traffic is increasing rapidly. In large part this is due to the establishment of commercially available data bases, which are expanding in scope and numbers. The data bases will be helpful in pure science; they are already proving very valuable in applied research, and industrial organizations are willing to pay well for tapping them. In fields such as chemistry, solid-state physics, and metallurgy, international competition is arising among compilers and vendors. We are in the early phase of important changes in electronic data handling. The federal government has been moderately helpful in furthering these developments, but expansion of its support is in order.

29 April 1983

Rapid progress is being made in research, development, and applications related to glass-fiber communication systems. A number of companies are active in this field, but at present the Bell System, including Bell Laboratories, is the leader. In February 1983, the Washington–New York link of its Northeast Corridor Lightwave Communication System became operational. Shortly thereafter, the Sacramento–San Jose link of a Sacramento–San Diego system was placed in service. The remaining links in both systems are under construction and will be completed in the next two years. The communication capabilities of the links in service are impressive, but recent disclosures of the progress of R&D indicate that the capabilities of fiber systems can be improved by orders of magnitude.

The Washington–New York link contains 30,000 miles of tiny glass fibers. Communication through the fibers is by means of light pulses generated by lasers that can turn on and off 90 million times a second. The fibers, consisting mainly of pure SiO_2, transmit light with a wavelength of 1.3 micrometers. In repeater stations every 4 miles, light is converted to electric pulses that are amplified and then fed to lasers for injection of pulses into the next section of the line.

One measure of the capabilities of the existing systems is that a pair of fibers could transmit the entire *Encyclopaedia Britannica* in 1 minute. But recent announcements make it evident that greater achievements will be forthcoming as a result of improvements in the fibers, and especially in the lasers. The starting material for the fiber is a thick-walled tube of silica glass. During the manufacturing process, the tube is rotated while being heated. A mixture of $SiCl_4$, $GeCl_4$, and O_2 is introduced into the hot tube and SiO_2 and GeO_2 are formed and deposited on the inner surface of the silica. With further heating, the tube collapses into a solid rod. Later, the rod is pulled out into fibers with an outer diameter of 100 micrometers. The inner core, with a diameter of 5 to 10 micrometers, has an index of refraction greater than the remainder of the fiber. Light transmitted along the fiber is confined to the core.

A key breakthrough has been the development at Bell Laboratories of lasers that emit light that is practically monochromatic and free of side bands. The speed of light is dependent on the wavelength. Confusing signals would reach the detector from a source having a number of lines. At a wavelength of 1.3 micrometers, the difference in velocity is minimal. Hence, the earlier systems employing lasers with side bands were designed to use that kind of light. However, the absorption curve of light in silica has a minimum at about 1.55 micrometers that is considerably lower than the minimum at 1.3 micrometers. Development of monochromatic lasers at 1.55 micrometers has opened that part of the spectrum for use in transmission. With changes in the composition of the gallium-indium-arsenic-phosphorus lasers, the emitted wavelength can be controlled. In principle, it will be possible to transmit simultaneously hundreds of independent signals through a fiber. At the same time, the attenuation of the light with distance has decreased. This makes possible a much longer interval between repeater stations. Improvements in the lasers also make possible the use of shorter pulses. In one experiment conducted by Bell Laboratories, signals were successfully transmitted at the rate of 420 million per second through 73 miles of fiber with no intervening repeater stations.

Most of the equipment developed for long-distance transmission is applicable to local networks. Glass-fiber loops have already been installed in a number of cities, including Los Angeles. With further R&D, costs will inevitably drop, and it will become practical to provide voice, data, and video services to offices, shops, and homes. Ian Ross, president of Bell Laboratories, has predicted that ultimately glass fibers carrying high-speed digital signals will make interactive video in color as widespread as today's telephone service.

23 July 1971

Science and technology have provided society with innumerable options and the ability to influence evolution. Optimists see in this a great future, with enhancement of the quality of life and of the dignity of mankind. Pessimists see quite a different picture, and at present they appear to be the more numerous and influential. The average citizen, when he or she thinks about it, is uncomfortable with the necessity of judging complex issues without adequate facts or background. The same individual also feels relatively powerless to affect the outcome.

In spite of the widespread feeling of ineffectiveness, some people have had very great influence and, collectively, the electorate is having profound effects. Public concern about misuse of technology is leading to measures designed to cope with such present abuses as pollution. Technology can be geared to ameliorate part of the disagreeable conditions, and constructive steps are now being taken.

Some of the difficulties created by science and technology are not so close to solution, particularly in biology and medicine. Advances in these fields have led to great benefits and also to puzzling problems, including some for which our present ethical concepts do not prepare us. More technology alone is hardly likely to provide satisfactory answers to the population explosion. Especially disturbing are aspects of the measures taken to prolong life in the very sick and very old. Death of a loved one was bad enough when it was in the hands of God; now it is often a much more distressing experience. Since every individual must participate in birth and death, he or she cannot escape some thought about the associated problems that science has created; in general, the individual is not grateful for the necessity to face such issues.

During the last several years, the public has repeatedly been warned that science is creating additional problems through raising the possibility of test-tube babies and "genetic engineering." The response of the public has been negative, with some

Anxiety about Genetic Engineering

calling for a halt to research in molecular biology. In truth, the dire predictions of the potentialities of new science have outrun the accomplishments, and the predictors have assumed that society will exercise negatively the options that are provided. Speculation about test-tube babies is based on a modest accomplishment—that is, fertilizing a human egg in vitro and keeping it alive for a week or so. For many years, biologists have been fertilizing eggs of countless species in vitro. Talk of genetic engineering received impetus from the isolation of an operon, a specific piece of DNA. This accomplishment is about as meaningful as the isolation of other components of the living system. Biochemists are experts at taking life apart, and they can reassemble some subsystems. The total system, however, is orders of magnitude more complex than anything they have put together. Even if biochemists achieve a capacity for genetic engineering, it is unlikely that their tools will match the tools that are already available. For example, artificial insemination is widely used to improve livestock. If some future ruling clique decided to engage in human genetic improvement, they would be more likely to adopt this technique and to employ their own semen than to use material concocted in the laboratory.

Talk of the dire social implications of laboratory-related genetic engineering is premature and unrealistic. It disturbs the public unnecessarily and could lead to harmful restrictions on all scientific research.

ENERGY AND THE ENVIRONMENT

Of all those areas in which science and technology intersect the world at large, none has so engaged Philip Abelson's attention or inspired his pen as energy and environmental topics—subjects that emerged as major public concerns during his editorship. The selection in this chapter exemplifies the knowledge and independence of vision that he brought to the task.

Consider, for example, his call for national cooperation at the height of the energy crisis, "Let the Bastards Freeze in the Dark." It commanded national attention and was widely quoted and reprinted, and not only for its forthright language; Abelson saw what much of the media had overlooked—that parochial attitudes toward energy were widespread and could be found in the environmental attitudes of the Northeast as much as in the energy selfishness of the South.

Abelson saw early on the potential of nuclear power and was initially enthusiastic about its ability to compete with coal as an energy source. But he also saw early—before the 1970s, before questions about safety became a major concern—the signs of economic difficulties in delays and escalating costs that have now all but stopped the expansion of domestic nuclear power.

Abelson has never minced words, even when, as sometimes happened, they were profoundly unpopular with much of his audience. In his look at nuclear fusion, for example, he called into question the high priority attached to research in this area, to the dismay of the physics community. He pointed out that practical power plants might be a long time coming and warned against neglecting far more prosaic energy research for the glamour of fusion. Likewise, he correctly foreshadowed (in the 1960s) the decline of domestic oil resources and our dependence on imports, and called repeatedly for more research on clean ways to burn coal, our largest energy source. And he has championed work on renewable sources of fuels, especially from biomass, pointing out their long-term significance to the world and their importance even now in the developing world.

At the peak of the DDT debate he reminded his readers of the benefits of that pesticide, while acknowledging that its use should be curtailed some three years before it was finally banned. He pointed out that water pollution problems were far more complex than just detergents, the focus of much of the public attention in the 1960s. He warned that cleaning up air pollution from automobile exhausts would be expensive—well before catalytic converters and emission standards—and asked for better documentation of the health risks involved.

Closer to the present is the complex scientific and political problem of acid rain. Here Abelson calls for national cooperation if workable solutions are to be found, given the enormous costs of cleanup, and counsels caution rather than haste, given our present lack of definitive knowledge. He points out that evidence of widespread damage to lakes is still lacking and that if damage to forests is the prime concern, we should focus on oxides of nitrogen (largely from automobiles) rather than oxides of sulfur (largely from power plants).

Fundamental to Abelson's approach to environmental questions has been the realization that these were profound problems that could not be solved overnight and that would require extensive scientific effort, major readjustment of public attitudes, and new economic arrangements. Thus, in a long list of pollution problems, his editorial voice is neither a shrill one demanding immediate solutions nor a petulant one denying the problem. Rather he sought the basic constraints—scientific, technological, economic—that underlie the problem. The result is a remarkably balanced point of view that on most topics has stood the test of time so well as to seem unexceptional today.

16 November 1973

A combination of environmental concerns and energy shortages is fostering widespread beggar-your-neighbor attitudes. Everyone wants cheap, unlimited energy, but all are prepared to fight tooth and nail to protect the environment and to prevent the location of energy facilities in their vicinity. It so happens that we are going to have neither cheap energy nor a perfect environment. Moreover, if we do not learn how to think nationally rather than parochially and to balance energy needs against environmental concerns, we are headed for trouble. Likely candidates for experiencing miseries are the people of the northern and eastern states. They have enjoyed cheap hydrocarbons from Texas, Louisiana, and Oklahoma. These supplies seemed to be guaranteed, but two developments have cast doubt on the security of supplies, particularly of natural gas. One is depletion of reserves. The other is a threat of indefinite dimensions. The southern states have awakened to the fact that they face future shortages and are looking for means of reserving their hydrocarbons for themselves. Bumper stickers can be seen in Texas, Louisiana, and Oklahoma with the motto, "Let the Bastards Freeze in the Dark."

Some Southerners were shocked recently by news from San Antonio. An electric power plant there will be fueled not by Texas oil or gas, but by coal, probably from Colorado. Louisiana, the principal supplier of natural gas to the eastern states, cannot expand industry as it would like because its gas is already committed. Moreover, some of its citizens will lose their jobs this winter because interruptible supplies previously available to Louisiana industry will be going north. Of the large curtailments of natural gas this winter, nearly half will be in the Gulf Coast region. Small wonder that the governors of Louisiana, Texas, Mississippi, and Oklahoma have joined in threatening to use police power if necessary to preserve their states' fuel supplies.

Governor Edwin Edwards of Louisiana recently pointed

out that people in other states are content to have oil gas wells drilled off the Gulf Coast and to have refineries running in his state and pipelines crisscrossing its terrain. "But these states don't want any of that activity in their own backyards," the governor noted, and then went on to say, "We're not going to be forced to continue operating our refineries and drilling new wells to deplete our resources in Louisiana to benefit the rest of the country."

The governor has a point. When Louisiana's resources are exhausted, who will furnish his people hydrocarbons, and at what cost? The governor was also on target in criticizing the attitude of the Atlantic Coast states toward exploration of their offshore resources. To the east on the continental shelf are thick sedimentary layers with substantial potential for oil and gas. Not a single exploratory hole has been drilled, however, and when the possibility of such exploration was mentioned about a year ago, there was immediate, strong opposition. Even if exploratory wells could be drilled within the next year, it would be years before natural gas would be flowing into homes. But events on the energy front have been moving with a time scale of weeks rather than years. It is not clear that the southern governors can curtail northern supplies of gas soon, but they will be under increasing pressure to do so.

Given resolute action and national cooperation, we can make it through the coming crisis without suffering. But a continuation of present energy consumption patterns, coupled with the trend toward parochial self-interest, can only lead to divisive tensions and some freezing in the dark.

14 December 1962

It is a curious aspect of human nature that we are more likely to be excited by unproved possibilities than by achievements. This is exemplified by the public response to a report that the Atomic Energy Commission has just released. This document has attracted little attention, yet buried in its pages is evidence of one of the great events of all time—emergence of nuclear energy as a practical source of power and a means of conserving fossil fuels.

Immediately after World War II a rash of forecasts extolled the wonders of the coming atomic age. Optimistic pronouncements of scientists were amplified by the press. There followed a period in which public interest remained high, though progress toward production of economically competitive electric power was slow, partly because competence in reactor design was vested mainly in physicists without engineering experience. Ultimately this deficiency was corrected, and a large number of technical problems were overcome. Different types of reactors were tested; new materials were investigated; every facet of fuel handling and safety was studied. At the same time a virile, competitive atomic energy industry was created.

Trends in the cost of generating electricity are one measure of what has occurred. These are outlined in the report.

> ... costs have been reduced, from the first actual experience of about 50 mills per kwh at the Shippingport prototype reactor in 1958 to less than 10 mills per kwh for full-scale plants now in existence and an estimated 5.5 to 6 mills for a large plant to be built in the near future at Bodega Bay, California.

This figure includes all such costs as amortization and does not involve a subsidy. The Pacific Gas and Electric Company has chosen atomic energy over fossil fuel not because of glamor but because of economics. Costs based on substantial operating experience now can be estimated with precision.

> ... the total nuclear electric generating capacity in the country [is] approximately 850,000 kilowatts, about 0.5% of our

Civilian Nuclear Power

> total installed capacity. Seven other central station nuclear power plants are scheduled to start operation in the next few months.

Economic civilian nuclear power has been achieved at a comparatively low cost.

> To date, the Commission has spent approximately $1.275 billion specifically on the civilian power program. . . . The present annual rate of expenditure is approximately $200 million.

One of the aspects that the report underplays is our vanishing resources of petroleum and natural gas. A chemical industry based on petroleum resources is increasing rapidly, and its products are assuming an ever-wider role in our economy and in international trade. With current trends we might be a have-not nation in this important sphere in about twenty years, with corresponding weakening of our competitive position. It is good to know that recent developments in nuclear energy have justified the optimism of fifteen years ago and that means are available for sparing our heritage of fossil fuels.

6 November 1964

The intense economic competition presently existing between conventional and nuclear power is bringing benefits to this nation on a scale that in the next decade may exceed a billion dollars a year. The pace of technological advance in power generation has been especially fast during the past two years. In this period the cost of generating nuclear power has been cut drastically and the cost of producing conventional power has also diminished.

It is instructive to analyze statements on the matter by Philip Sporn, a leading spokesman of the electrical industry. In 1962, when Sporn estimated future costs of nuclear and conventional power, he flatly stated that nuclear power was not competitive with conventional energy. He also estimated that in the period from 1973 to 1978 nuclear power would cost between 6.17 and 6.89 mills per kilowatt-hour, whereas costs of conventional power would vary from 3.9 mills in favorable areas to 5.6 mills in high-cost fuel zones. Thus, in 1962 a leading expert considered that even after fifteen more years nuclear power would not be competitive.

In two years the outlook has changed surprisingly. The General Electric Company has entered into a contract to build a nuclear installation at Oyster Creek in New Jersey. This plant, to be completed between 1967 and 1968, is expected to deliver power at a cost as low as 3.66 mills per kilowatt-hour. Sporn has prepared a new analysis of the competitive status of conventional and nuclear power, and his views are different from those of two years ago. He is unwilling to accept the Oyster Creek plant costs as typical, contending that the General Electric Company has priced its plant too low, for competitive reasons, and perhaps has been too bold in guaranteeing performance. Even so, he concedes that there has been "an impressive . . . reduction in total energy costs" for nuclear power during the past two years.

In his report, Sporn also emphasizes the continuing improvements in the conventional approach to power, and he credits these, at least in part, to the competition between the two major sources of energy. In his own company, a plant using low-priced coal, to be completed in 1967, is expected to deliver power at 3.59 mills per kilowatt-hour, a cost below his earlier estimate of what might be achieved in the period from 1973 to 1978.

Perhaps the most impressive feature of Sporn's analysis is the change in his view of the energy competition. He now believes that this competition has reached the stage where nuclear power "is capable of joining this battle armed only with its own remarkable record of achievement and the promise of advancing further the established record of cost and performance without justification for, or need of, federal assistance."

It appears that another federal review of the energy situation is inevitable and may occur during the next session of Congress. Up to the last year or so, subsidies have been necessary to enable nuclear energy to compete. In the light-water reactor field that need is no longer clear, since the Oyster Creek plant will be constructed without direct government support. Nevertheless, it is to be hoped that Congress will move cautiously in changing the rules of the game. We are witnessing a beneficial competition that should not be interfered with lightly. In addition, development of breeder or advanced converter reactors should be given high priority. Success in this effort would have tremendous long-term significance. Our nuclear scientists and engineers should be given every encouragement and incentive to maintain their record of superlative achievement.

12 July 1968

A dramatic confrontation between rosy optimism and harsh reality is now gripping the attention of the electrical power industry. During 1966 and 1967, in a bandwagon atmosphere, large numbers of nuclear power plants were authorized. Several months ago, about 35 percent of scheduled additions to electrical capacity were nuclear. Recent events, however, have caused some observers to fear that optimism was overdone. The utilities have gambled heavily on unproven equipment, some of which will be brought on-line far behind schedule. Power shortages could result.

A conspicuous example is the installation at Oyster Creek, New Jersey, which is now about a year and a half behind schedule. During field hydrostatic testing of the reactor pressure vessel on 29 September 1967, a leak was detected. A dye-penetrant test revealed that the leak was the result of flaws in a field weld made to join a control rod housing to a stub tube in the pressure vessel. Detailed examination revealed localized intergranular cracking in 123 of 137 stainless steel stub tubes, and welding defects in each of the 137 field welds joining the stub tubes and the control rod housings.

Many of the defects found were minor, and it seems unlikely that complete failure of a weld would have occurred had the weaknesses not been discovered. Even had such failure occurred, there would not have been a violent nuclear accident. However, if a leak or a weld failure had occurred after the reactor had operated for some time, the difficulty of repairing the defect would have been great, owing to intense radioactivity.

Before the Oyster Creek facility can be operated, it must be licensed. Three different groups will pass on the matter. First, there is the Division of Reactor Licensing of the Atomic Energy Commission, then the statutorily constituted Advisory Committee on Reactor Safeguards, and finally the Atomic Energy Commission itself. These bodies cannot be expected to act hastily. Defects in one aspect of the plant raise specters of other, yet

Nuclear Power: Rosy Optimism and Harsh Reality

undetected, flaws, and it is not certain that procedures used for repair of the defects will be acceptable. When the Oyster Creek generating plant will become operational is anybody's guess, but it could be in the distant future.

These delays will be costly in money and prestige. The Oyster Creek plant represented a courageous gamble by the General Electric Company, which, in 1963, undertook to guarantee delivery of a completed plant involving new design features at a stunningly low price. Announcement of the contract for the plant was widely regarded as signifying that nuclear power had come of age.

Following this event, other large nuclear installations were authorized at an increasing rate. Then came a great outcry against air pollution associated with coal-fired plants. The move toward nuclear power became a stampede. Delays at the bellwether Oyster Creek plant will have a sobering effect. An additional deterrent is the fact that costs of nuclear installations have increased by 40 percent during the last two years. Nuclear plants also have been tagged as important potential contributors to thermal pollution, since they are relatively less efficient thermally than coal-fired plants.

All of these difficulties will be surmounted, and nuclear power one day will furnish a substantial fraction of this country's electrical energy. How distant that day will be will depend mainly on how long it takes industry and labor to achieve new and higher standards of design excellence and quality control.

23 July 1976

Viewed from a distance, energy from the fusion of light nuclei is a glamorous concept. Advocates have talked of obtaining unlimited amounts of cheap, clean energy from the virtually inexhaustible deuterium of the oceans. But looked at closely, deuterium fusion is far from practical application; if achieved it will be costly, and it will create large quantities of radioactive substances.

The radioactivity associated with fusion arises when neutrons produced by the reaction of deuterium with deuterium or tritium collide with wall and structural materials. In the generation of a given amount of energy considerably more neutrons are produced by fusion than by fission. This is true whether the fusion occurs in a magnetically confined plasma or is induced by lasers. Thus, shortly after a fusion device begins to generate energy in substantial amounts, the inner walls will become so radioactive as to prevent direct contact by workmen. With more lengthy exposures a radioactive waste disposal problem will ensue.

Neutrons undergo many nuclear collisions before they are absorbed. These encounters damage the crystalline structure of solids. In a practical fusion device each atom of the containing inner wall would be drastically displaced many times a year. Moreover, the very fast neutrons from deuterium-tritium reactions create hydrogen and helium within lattice structures, leading to embrittlement. These damaging radiation effects would be likely to necessitate replacement of highly radioactive materials, requiring the use of complicated remotely controlled equipment.

To obtain a useful rate of fusion in the easiest case—deuterium plus tritium—energies of 10 kiloelectron volts, corresponding to 100,000,000°C, are required. The fusion device that is currently receiving most funding is the tokamak. This machine employs magnetic confinement to keep the hot plasma away from metal walls. To achieve production of practical amounts of energy, a very large volume of plasma and a correspondingly large magnetic field are required. But more than just any large magnetic field is needed and the problems of confining plasma of a useful density may never be solved. During the past twenty years, some of the world's most talented physicists have worked toward practical fusion. The outstanding lesson that they have learned is that hot plasmas are difficult to confine.

The U.S. contingent in fusion research has been supported with a total of about $800 million. Thus far the results, though moderately encouraging, are not impressive—to produce an output of 1 watt of fusion power requires an input of about 10^6 watts. To reach a break-even condition will apparently require building a succession of very large, very costly devices employing superconducting coils. The total cost of attaining a 10-megawatt output by a fusion unit in 1990 has been estimated to be $10 billion. This is in contrast to a cost of $100 million to reach the same energy output by a solar device or a breeder reactor.

Moreover, the tokamak device does not produce power continuously but must go through a cycle during which output falls to zero. Utility executives have enough headaches without trying to cope with that kind of source of power.

Nuclear fusion has been hailed as one of the world's three major long-term energy options along with breeders and solar energy. Ultimately, its great potential may be harnessed and this nation should continue to support efforts to reach the goal. However, thus far fusion has been like a pot of gold at the end of a rainbow. The beauty of the rainbow should not dazzle us into depending on riches we may never see. Nor should the glamour of the challenges of the high technology of fusion be allowed to keep attention away from such humdrum measures as conservation and the development of a practical photosynthetic energy source.

30 December 1966

Most of the energy needs of the United States are met by fossil fuels. In 1960, petroleum, natural gas, and related products supplied 71.8 percent of the energy consumed, while coal and lignite furnished 24.5 percent. Nuclear energy accounted for 0.1 percent.

This pattern will change. The sudden emergence of competitive nuclear energy has been noted. Not so evident are trends in the availability and use of petroleum and related products. At one time the United States was a low-cost producer and exporter of petroleum. Today we are not competitive and we import more than 20 percent of our needs at an annual cost approaching \$2 billion. Proved reserves of crude oil of 30×10^9 barrels are adequate for the next decade, and more oil will be found. However, the cost of finding oil is increasing, and the United States is no longer an attractive area for petroleum exploration. The most important petroleum discoveries are being made in the Middle East and Africa.

There is controversy over estimates of the amounts of oil likely to be found in the United States, despite the fact that 1.9 billion feet of exploratory holes have been drilled. M. King Hubbert has shown that present-day drilling is discovering less oil per foot of drilling than was found a generation ago. He estimates that, during the period from 1930 to 1935, each foot of hole drilled led to discovery of an average of 160 barrels of oil. During the period from 1960 to 1965, he estimated that 39 barrels of oil were found per foot drilled. This comparison includes an estimate of the amount of oil to be added by development drilling and extensions.

Recently Robert O. Anderson, chairman of Atlantic Richfield Company, implicitly confirmed the view that prospects are not good for discovering petroleum in the United States. He warned that, by 1976, imports of petroleum could account for a third of total U.S. consumption, at a cost of \$4 to \$5 billion annually. This would aggravate an already serious problem in balance of payments and would place us at the mercy of Middle Eastern and other foreign suppliers.

Fortunately there are alternatives. The oil-shale deposits of the western United States could be exploited to produce more than 10^{12} barrels of oil. Hydrogenation of coal also could supply enormous quantities of liquid fuels. Government and industry have not made great efforts to develop these alternative sources. Estimated expenditures for energy research and development during 1963 for government and industry, respectively, were (in millions of dollars): coal, 11 and 11; oil and gas, 40 and 336; and nuclear fission, 210 and 90.

Research and development work sponsored by the Office of Coal Research of the Department of the Interior looks promising. Notable are two developments being conducted by Consolidation Coal, now a division of Continental Oil Company. One involves solvent extraction of coal followed by hydrogenation to produce liquid fuels similar to gasoline. A second development is an improved process for synthesizing methane from lignite.

This nation has been slow to respond to a deteriorating position in petroleum exploration. There should be substantially increased effort to develop substitute supplies through research, development, and economic incentives.

22 January 1982

Coal seems destined to become a major source of energy and chemicals. But present technology for its use is primitive. Direct combustion of coal results in the formation of oxides of sulfur and nitrogen. These gases have been linked to acid rain, which has become an increasing source of concern. Flue gas desulfurization is being practiced in new power plants and removes as much as 90 percent of their sulfur oxides, but this is accomplished at the cost of reliable operation of power plants and the production of enormous amounts of calcium sulfate sludge.

Improved technology for utilization of coal is on its way. Of special significance are developments in gasification. By this means one can quantitatively remove in the form of hydrogen sulfide and ammonia the sulfur and nitrogen that were originally present in the solid. When the gasification is accomplished at a high temperature with oxygen and water, the coal is completely destroyed and synthesis gas (carbon monoxide plus hydrogen) is formed. This gas can be burned as a fuel, processed to yield methane, or used as a feedstock for synthesis of many chemicals.

A number of processes have been devised for gasification. These include the Lurgi, the Koppers-Totzek, and the Texaco processes. It seems likely that the Texaco process will become widely applied. Its principal feedstock is a coal slurry, and coal slurry transportation will have a large future role. The process is flexible in its utilization of a wide variety of coals. The gasifier operates at temperatures of 1250° to 1500°C and pressures of 350 to 1200 pounds per square inch. Under these conditions the organic matter in the coal is completely gasified. There are no tarry residues. The Texaco process is highly suitable for integration with combined cycle electric power generation. This application is now about to be tested in a $300 million installation being built in southern California. It is designed to convert efficiently the thermal and chemical energy of the hot gases into electricity. In order to remove particulate matter and H_2S and

NH_3 from the gas, it must be cooled. This will be done by using heat exchangers, from which hot water and steam will be obtained. The steam will drive a turbine. Once the gas has been cleaned it will be burned in a gas turbine. Heat from the hot combustion gas will also be used to make steam.

The demonstration plant is designed to use 1000 tons of coal a day and to produce 100 megawatts of power. It is scheduled to become operational in two years. Many tests will be conducted on it. Different coals will be burned. Its performance as one of the components of a generating system will be studied. Of particular interest will be the response of the plant to sudden changes in electrical demand. One of the hoped-for advantages of this type of plant is that it will be cost-efficient on the scale of 100 to 300 megawatts. It is further believed that such plants could be constructed more rapidly than the 500- to 1000-megawatt installations currently being built. This would give the utility companies much more flexibility in matching capacity to unpredictable demand.

The demonstration plant is being constructed under the sponsorship of the Electric Power Research Institute, Southern California Edison, General Electric, Bechtel, and Texaco. No federal funds are involved. Siting of the plant in California makes good sense. As much as 79 percent of electricity generated by Southern California Edison is derived from oil or natural gas. The plant will easily meet the rigorous state standards for air pollution.

Sponsors of the demonstration plant merit congratulations for their initiative and risk-taking. It is likely that they will be rewarded. This new venture shows great promise, and its success would solve many of the problems of the utilities while giving them an opportunity to have a role in the synthetic chemicals field.

---

26 March 1976

There has been much talk of solar energy, but most thought has been devoted to physical means of collecting sunlight. An obvious resource has had only nominal attention: energy, materials, and chemicals from plants and trees. For the coming decades in the United States, the major shift from oil and natural gas is likely to be toward coal and oil shale. In many countries, however, where there is little or no coal or oil shale, trees and plants could become important sources of energy and materials. Moreover, burning of fossil fuels yields carbon dioxide—a hazard of unknown magnitude. The use of trees and plants as sources of energy involves a closed cycle with respect to carbon dioxide.

In comparison to the amounts of incident solar energy on earth, human utilization of energy is relatively trivial ($\sim 5 \times 10^{-5}$). The people of the United States consume an amount that is about 10^{-3} that of solar energy falling on it. Under favorable conditions, about 3 percent of solar energy can be fixed in photosynthesis. Thus, in principle, energy needs of the United States could ultimately be met by devoting only a small fraction of the land to this purpose. The same is true of the rest of the world. The present annual production of biomass on the land areas has been estimated at 100 billion tons (dry weight). This has an energy equivalent that is about a factor of six greater than current utilization of energy by the world's peoples. Moreover, these figures do not take into account the large additional production of biomass that might be achieved through agronomy and silviculture.

There is, of course, a long distance between potential and practice. Feeding the earth's billions of people has top priority though only a small fraction of the energy is represented by food. Perhaps a more restraining factor is the diffuse character of the biomass. Effective utilization of the potential might require many small conversion plants. This would facilitate return of nutrients to the land while cutting costs of transportation. But at present prices and with current technology, only a tiny portion of energy needs in the United States could be met today from renewable sources.

Energy from Biomass

Ultimately, trees may be the preferred source of energy. However, effective use of them for other than direct burning involves complex technology. Cellulose and lignin must be separated and then processed further or the wood must be converted to carbon monoxide and hydrogen and from thence to methanol or methane. Plant materials lend themselves to anaerobic fermentation resulting in methane and carbon dioxide with the latter easily removed.

One potentiality that merits further study is the hydrogenation of wood that is known to yield combustible liquids.

The United States has so many possible energy sources and so much wealth that it can perhaps afford long controversies about what to do. Other nations cannot equivocate. They will adopt nuclear energy unless alternatives quickly become available. This will be true ultimately even in the tropical, less-developed countries that have abundant renewable resources. Where will those countries find the billion or so dollars to buy a nuclear power plant? Where will they find the skilled technicians to operate the plants safely? Would the world not be better off and safer if such countries were obtaining their energy from biomass?

Thus, although there seems to be no great urgency in the United States to develop renewable energy sources, it would be desirable to give this matter high priority and substantial funding. The effort should include fundamental work in photosynthesis, plant genetics, artificial creation of new species, and related aspects of agronomy and silviculture. Research and development work should include imaginative approaches to better processes for utilizing plants and trees. Some of the efforts should be devoted to creating simple inexpensive devices that would enable rural peoples to obtain various types of energy including even electricity from biomass. With a moderate investment of money and scientific and technical personnel, we could perform work of immense global significance while moving toward long-term solutions to our own energy problems.

---

9 May 1969

Controversy over pesticides is heating up. Michigan and Arizona have banned the use of DDT and, beginning in 1970, Sweden will restrict its application. Health, Education and Welfare Secretary Robert Finch has appointed a commission to tell the government what to do about it, and the Food and Drug Administration has seized shipments of coho salmon containing DDT or its derivatives at concentrations of more than 5 parts per million.

Few man-made chemicals have had as profound an effect on human lives as DDT has had. It has been instrumental in saving many millions of lives, in part through control of insect vectors, in part through increased food production. The quality of many crops has been improved greatly through its use, and in some instances yields have more than doubled. Were the United States to stop the use of all pesticides, food prices would rise sharply and supplies of many foods would be inadequate. Although some people might prefer banning all pesticides, the real issues are the choice of pesticides (for example, persistent or nonpersistent) and the conditions under which they are to be used. In this controversy, some agriculturalists and industrial interests defend the application of persistent chemicals such as the chlorinated hydrocarbons—DDT, dieldrin, endrin, and others. Conservationists and many members of the concerned public advocate the use of nonpersistent chemicals such as the carbamates or the phospho-organics—for example, malathion. The defenders of persistent chemicals point to their effectiveness and low cost. DDT is a relatively inexpensive chemical, and one or two applications may suffice for a season. Much of the cost of using pesticides arises from the labor involved in applying them. Hence, complete abandonment of the chlorinated hydrocarbons now would result in a substantial increase in the nation's food bill.

The defenders of DDT point to the remarkable fact that in three decades of use there has been no documented instance in

Persistent Pesticides

which human deaths have resulted from proper application of the chemical, and relatively few deaths have occurred even with drastically improper use. Moreover, DDT and its relatives are not truly persistent but are slowly destroyed in soil. DDT is slowly degraded in man, and it is also excreted, so that concentrations do not build up indefinitely. Typical human fat contains DDT and its degradation products at concentrations on the order of 12 parts per million.

An undesirable property of the chlorinated hydrocarbons is that they are somewhat volatile and may be carried far from the point of application. Residues of DDT have been found in seals and penguins in Antarctica. When a farmer in Europe applies the chemical to crops, some of the molecules are destined to accumulate in you and me.

Part of the current concern about DDT is due to new biological findings in animals other than man. Among the effects demonstrated have been an interference with shell deposition in some birds and an estrogenic stimulus in rats by a component of commercial DDT. Highly controversial is a report, soon to be released, that describes carcinogenic effects of large amounts of DDT in tumor-susceptible mice. Most scientists would agree that use of DDT should be curtailed, and indeed consumption of DDT in the United States has been declining; during the 1967 crop season only half as much was applied as in the peak year of 1959. However, use in other countries is increasing.

Suppression of the application of chlorinated hydrocarbons will not end controversy over pesticides. Unanticipated, possibly dangerous, side effects of the new substitutes will be discovered. Conservationists will continue to criticize, and that will be useful, for all of us have a large stake in the quality of the environment.

20 May 1966

A few years ago the Wisconsin State Board of Health reported the existence on the Mississippi River of a "wall of foam 35 feet wide, 300 feet long and 15 feet high." This was only one of many photogenic newsmaking incidents. Consequently, the issue of water pollution was publicly dramatized and oversimplified. Synthetic detergents were spotlighted as sources of the nuisance. The principal chemicals involved were alkyl benzenesulfonates, the alkyl group usually being a C_{12} branched-chain hydrocarbon. Biological degradation of this structure is slow. Threatened with congressional action, the detergent manufacturers abandoned production of the branched compounds, using instead a linear alkyl benzenesulfonate that is degraded more readily. This changeover (made at a cost of $150 million) was completed by June 1965. In hearings last week before a committee headed by Senator Edmund Muskie, a year of field experience with the new products was described. The input and output of detergents from several major sewage treatment plants have been monitored. As a result, it was possible to follow closely the consequences of the changeover. Part of the chemicals now are degraded on the way to the treatment plant, and the rest are more easily removed once there. When 90 percent of the other sewage is consumed, a similar fraction of the new detergent disappears.

In spite of this excellent experience with modern treatment plants, the manufacturers may yet face further pressures. Much of the sewage in this country is disposed of in cesspools or septic tanks. In well-designed systems the effluent flows into surrounding soil, where it is acted on by aerobic bacteria, and the new detergents are removed. However, if free oxygen is absent, the detergents are not degraded. Householders using wells may still find their water foamy.

The dramatic aspects of the detergent problem, however, seem solved, and this is good. Now our people and the Congress can turn to the substantive problems of water pollution. Only a minor fraction of the sewage from towns and cities is

Water Pollution

fully treated. Upstream communities show little enthusiasm for spending money for the benefit of communities farther down. Even in those cities that have "full treatment," performance is often poor. In most instances the storm drains and the sewage systems are interconnected. When a storm occurs, the treatment plant is bypassed. This occurs at a time when the scouring action of high-velocity flow dislodges large quantities of solids that have accumulated during low flow. The problem will not be quickly disposed of. The cost of separating storm and sanitary sewers has been estimated at $20 to $40 billion.

Water pollution is not one problem but many. Each watershed requires a different approach, depending on technical, economic, and political factors. A particularly instructive case to follow will be events at Lake Tahoe. This beautiful lake is threatened with degradation. Once relatively poor in algal nutrients, the water has been fertilized by effluent from treated sewage. The communities surrounding the lake now plan to pump their effluent out of the Tahoe watershed. However, even this practice may not suffice. It may become necessary to ban use of fertilizers on garden plots around the lake.

The problems of water pollution are many and complex. The greatest present difficulty is that, while the public favors pollution abatement, only a few politicians are providing imaginative leadership.

26 March 1965

Air pollution is becoming a serious national problem. Formerly it was a local issue largely connected with industry, but today a principal source of increasing pollution is the ubiquitous automobile.

When motor vehicles burn fuel they produce a number of products in addition to carbon dioxide and water. Important amounts of carbon monoxide and nitrogen oxides are formed. The fuel is not entirely consumed. Part is exhausted unchanged, part appears as hydrocarbons of smaller molecular weight, including reactive olefins.

At a concentration of slightly more than 1000 parts per million carbon monoxide kills quickly. Most people experience dizziness, headache, and other symptoms at approximately 100 ppm. Concentrations as high as 72 ppm have been observed in Los Angeles, and values above 100 ppm have been measured in Detroit. In almost every metropolitan area peak concentrations of carbon monoxide approach the 100-ppm level.

In California efforts have been made to decrease the amounts of carbon monoxide emitted by motor vehicles by use of devices such as catalytic afterburners. At the same time there has been a trend toward higher combustion-chamber temperatures. These efforts result in more complete combustion but also contribute to an increase in the production of oxides of nitrogen. Nitrogen dioxide is a poisonous brown gas. The threshold level of toxic effects is not well known, but it appears to be about 5 ppm. On one occasion a concentration of nitrogen oxides of nearly 4 ppm was observed in Los Angeles.

Automobile exhaust products interact to produce physiological and chemical effects that are greater than the sum of the parts. Synergistic effects of carbon monoxide and nitrogen oxides on respiration have been noted. Light hydrocarbons alone are not very toxic, but in the presence of nitrogen dioxide and sunlight, hydrocarbons give rise to noxious substances. Nitrogen dioxide acts as a photoreceptor and is decomposed to nitrogen oxide plus atomic oxygen. This reactive form of oxygen attacks hydrocarbons. The products may react with molecular oxygen to form peroxyl radicals. These in turn react with oxygen to form ozone. The oxidants react further with the original materials as well as with their reaction products. The result is a complex mixture of toxic substances.

As yet there is little evidence of chronic effects from air pollution. However, a large fraction of our population is now being exposed to significant concentrations of a variety of toxic chemicals. These levels are often a substantial fraction of those which produce acute effects. There is a possibility that our people may be sustaining cumulative insidious damage. If genetic injury were involved, the results could be especially serious.

At present we cannot accurately evaluate the hazards of air pollution. The toxicity of even some of the simple important chemicals is not well established. It is clear that there are acute synergistic effects, but these have not been thoroughly examined. Even so basic a problem as establishment of good methods for measuring the concentrations of pollutants has not been completely dealt with.

The automobile and the automotive industry are central to our way of life and to our economy. Can we live with a constantly increasing level of pollution, or will we be forced to take drastic steps to protect the nation's health? At present the federal government is spending about $24 million a year on the study of all aspects of air pollution. Considering the potential seriousness of the problem, this sum seems much too small.

10 April 1970

The U.S.S.R. has assigned top priorities to armament, space, and capital goods. Only a few luxury consumer goods, such as caviar, are produced for export. But even this item is now disappearing, and its price has risen sharply.

Natural caviar is derived from the roe of the sturgeon. Most of these fish dwell in the Caspian Sea and the Volga River. Pollution of these waters has led to attenuation of the fisheries. A Novosti Press Agency release states:

> Off-shore drilling operations keep expanding, and, as a result of a purification bungle and an irresponsible approach to this problem by the oil-extracting plants and refineries, the Caspian Sea . . . is becoming badly polluted. Add the industrial sewage discharged into the sea and the contamination by intensive shipping, and you will not be surprised to learn that the sturgeon catch is falling sharply.

Supplementing these remarks is a passage in *Soviet Life*:

> The Caspian's prime polluter is oil. Until recently all the off-shore oil installations, which use great quantities of water, dumped the contaminated water into the sea. Hundreds of tankers carry oil derivatives across the Caspian. The holds of the tankers were cleaned en route, and the waste water emptied into the sea.

Examination of Soviet publications indicates that their water pollution problem is widespread and serious. Oils, phenols, alkalis, acids, and organic wastes are dumped in streams and lakes, and only a small fraction of plants have adequate facilities for waste treatment. A passage from *Izvestia* states:

> There are more than fifteen thousand milk, butter and cheese factories and separator departments in the country . . . they consume millions and millions of cubic meters of water. The number of fish factories, tanneries, linen factories, regional food combines and industrial complexes is still greater . . . nearly all of these enterprises have no waste water purifying installations.

The Soviet government is now moving toward abating pollution, but the problem will not be solved quickly. Huge capital investments in treatment facilities are required, and construction will extend over many years. The Soviet government must also cope with its plant managers. Another quotation from *Izvestia* illustrates the problem. In a discussion of the failure of management to construct purification installations, the item says:

> The Voskresensky Chemical Combine managed not to spend one kopek on this construction, although the money had been allocated. They explain how, waving their arms, the design organization did not turn over the drawings on schedule. But why did the combine director, Comrade Doktorov, instead of trying to obtain the designs, begin to fuss about to have himself relieved of all these unnecessary headaches, the installations of all kinds of filters and sediment traps?

In commenting on this quotation, Myron Tribus pointed out that the problem of managing pollution is "universal, characteristic of all technological societies and in the end reflects the value judgments of those people who are creating the wastes." Senator Edmund Muskie summarized the matter in another way when, in speaking of the Russian and American approaches to pollution, he said, "It is not so important who owns the means of production as how they are managed."

Both the U.S.S.R. and the United States have been careless in despoiling the environment. Both now seem to be moving toward a more responsible posture. If the two nations were to compete in the cleanup process, that would be constructive. If they were to cooperate, that would be a welcome miracle.

17 July 1970

During this century, in the United States, about 75 million kilograms of mercury have been consumed; little information is available on its final disposition or on the concentration of the element at specific points in the environment. Recently, it has become clear that compounds of mercury present a substantial hazard. Of particular significance is methyl mercury, a highly toxic substance that causes neurological damage, produces chromosomal aberrations, and has teratogenic effects. It is mainly in this form that mercury is found in food fishes. Recent studies have elucidated some of the steps in the accumulation.

Industrial wastes containing inorganic mercury or phenyl mercury find their way into bottom muds of lakes. There they are converted by anaerobes into CH_3Hg^+ or $(CH_3)_2Hg$. The latter compound is volatile, and it escapes into the water column from the sediment. Though quite stable in alkaline solutions, $(CH_3)_2Hg$ is converted to CH_3Hg^+ at low pH. This ion is soluble in water, and it is concentrated by living things, usually appearing in the body lipids. In part, the concentration may come by way of the food chain, but apparently fishes may also accumulate the toxic ion directly. The concentration factor from water to pike is of the order of 3000 or more.

Substantial mercury pollution in the Great Lakes became apparent in March of this year. Mercury concentrations as high as 5 parts per million were reported in some pickerel shipped from Canada. Typical concentrations in fish taken from Lake Erie were 1 ppm or less. Further investigations have confirmed the existence of a major environmental problem traceable to the dumping of large amounts of mercury-containing liquid wastes. This discovery comes as a surprise to most scientists and apparently to federal authorities. However, there was ample reason for looking for such a phenomenon. Episodes in Sweden and Japan had pointed to dangers arising when large quantities of mercury are discharged into the environment. In Sweden the use of methyl mercury in a seed dressing had led to a drastic

Methyl Mercury

decrease in wild bird populations. Near Minamata, Japan, between 1953 and 1960, 111 persons were reported to have been killed or to have suffered serious neurological damage as a result of eating fish and shellfish caught in mercury-contaminated areas. Among the 111 were 19 congenitally defective babies born of mothers who had eaten the fish and shellfish. Later at Niigata, Japan, 26 cases of methyl mercury poisoning were noted. The affected persons and their families ate fish with a frequency of 0.5 time to 3 times a day; the fish contained mercury at concentrations of about 5 to 20 ppm.

Physiological and cytological studies have revealed some of the behavior of methyl mercury. It tends to be associated with red blood cells and nervous tissue, and it easily passes the placental barrier, becoming moderately concentrated in the fetus. It can cause chromosomal disorders. Fruit flies consuming food containing methyl mercury at a concentration of 0.25 ppm had offspring carrying one extra chromosome.

It seems unlikely that anything approaching the Japanese observations will be seen in the region of the Great Lakes. There the concentrations of methyl mercury in fish are lower and fish are a less important part of the diet. Nevertheless, we have a substantial and long-enduring problem; even if fresh pollution were stopped, it would be many years before natural processes could cleanse the lakes.

This episode has not led to demonstrable tragedy, but it should remind us how much we risk when we convert our rivers to sewers and our lakes to cesspools.

18 May 1984

Nearly 50 percent of the people in the United States are dependent on groundwater for their supplies of drinking water. Both nature and humans act to contaminate this essential resource, and public concern about its quality has been building. Recently, two publications [V.I. Pye, R. Patrick, J. Quarles, *Groundwater Contamination in the United States* (University of Pennsylvania Press, 1983) and *Studies in Geophysics: Groundwater Contamination* (National Academy Press, 1984)] have appeared that describe the extent of such pollution, indicate scientific considerations, outline steps to be taken, and note laws bearing on the matter.

The sources of pollution are many, and they differ from place to place. Industrial wastes have received most of the publicity, but other sources are of comparable importance. Agriculturally related contamination in the western United States is extensive and is likely to become increasingly serious. Municipal landfills let loose a complex collection of pollutants. Effluents from septic tanks on Long Island contaminate the drinking water of some of the three million people living there.

The total effect from all sources is to pollute badly about 1 percent of the aquifers. Because many of these aquifers are close to large population centers, the impact is disproportionately large.

The various contaminants interact with the background environment in different ways. In the aerobic zone, bacteria oxidize many of the organic constituents. But, in general, hydrocarbons are not metabolized under anaerobic conditions. In contrast, chlorinated organic compounds seem to be more likely to be attacked in anaerobic environments than under aerobic ones; chlorine atoms are removed.

The underground environment has considerable binding tendencies. For example, where present, zeolite compounds have an ion exchange binding capacity that is effective in holding cationic forms of toxic heavy elements. Most sedimentary horizons contain some organic matter to which hydrophobic organic molecules tend to be adsorbed. Thus, depending on the

Groundwater Contamination

path that groundwater traverses, some of its contaminants may be removed.

The rate of motion of the fluid is related to the permeability of the material through which it flows. A typical rate is about a foot per day. However, differences in permeability of 15 orders of magnitude have been noted. In general, the deeper the horizon, the less the permeability. Motion in a fine-grained material like clay is much slower than in a coarse-grained sand.

Most drinking water comes from wells that are less than 100 meters deep. Thus in some areas, waste fluids with a density greater than 1.0 could be safely injected if emplaced below 100 meters. An even more secure disposal area is in the arid Basin and Range country of the West. In that region there are basins that have no outflow.

One of the lessons of the past is that careless disposal of wastes can lead to problems that cost billions of dollars to correct. An obvious method of avoiding future additional groundwater problems would be to stop pouring wastes into the ground. For example, combustion of organic wastes would change them to simple products. Proper design of waste lagoons can guarantee that little of toxic substances escape to the environment.

The two publications make it clear that we are only at the beginning of gaining knowledge about underground transformations of substances and about the motions of their carrying fluids. New sensitive analytical instrumentation will be crucial in identification and quantification of migrating species. Laboratory development of bacteria with metabolic capabilities for wastes may prove helpful. Mapping of the underground aquifers should make it possible to protect our sources of potable water.

The rate of contamination of groundwater appears to have slowed but residuals from earlier carelessness remain. Ultimately, we must move toward improved methods of dealing with waste problems. Better knowledge of the underground environment will be an essential element in that progress.

---

25 November 1983

Two recent reports have dealt with the climatic effects of increasing amounts of atmospheric carbon dioxide (CO_2). The tone of the reports is less panicky than that of earlier statements. This is particularly true of the study conducted under the auspices of the National Research Council (NRC) [*Changing Climate* (National Academy Press, 1983)]. Earlier predictions were based on the then current rate of increase in combustion of fossil fuels. This amounted to 4.3 percent per year and would have led to a doubling of the concentration of atmospheric CO_2 in about fifty-seven years, with an estimated rise in global temperature of 3.0°C. Estimates of average annual rates of increase in CO_2 emissions to 2030 range from 1 to 3.5 percent. Using a rate of 2.0 percent gives an estimated doubling time of eighty-eight years. Perhaps more important are some considerations about possible societal impacts. The NRC report pointed out that human societies are flexible in dealing with new situations provided sufficient time is available. As an example, the great changes that have occurred in this century were cited.

The NRC report was careful to outline uncertainties in the predictions. The possible temperature rise is based on model studies that may or may not be valid. The projected rise is necessarily to be superimposed on unpredictable natural climatic trends. More controllable, but still unpredictable, is the rate of burning of fossil fuels. For nearly two decades ending in 1973, consumption expanded at a compound rate of 4.3 percent per year. During the past decade the rate of burning has been static. If present trends continue, the doubling time for CO_2 will be 220 years.

Whatever the rate of increase of CO_2 content and corresponding change in temperature, ultimate melting of a large Antarctic ice mass seems highly probable. This would lead to an estimated rise in sea level of 5 to 6 meters and to flooding of highly populated areas. Can such an event be delayed or even forestalled? The answer is that it probably can. Continued effort to increase the efficiency of energy use could lessen demand. A number of measures could be employed to slow the rate of increase of CO_2. One method for decreasing net emission of CO_2 is close to commercial feasibility. It has the potential advantage of curtailing the emission both of CO_2 and of gases responsible for acid rain. In an electric power plant now being built at Cool Water, California, coal is gasified and impurities such as sulfur are captured. Following the combustion of fuel gases, the CO_2, being present in high concentration, could be easily removed. Later it would have a market value for injection underground to promote tertiary recovery of oil.

A transition to greater dependence on renewable energy would also be helpful. It is useful to be reminded that energy consumption by humans amounts to only 0.1 percent of the solar energy falling on the earth. Recent progress in tapping some of this energy by improving the increased production of biomass is large. With good management and superior choice of vegetation, CO_2 fixation might be increased fivefold or more. The product would be sufficient to sustain a prosperous civilization. Any surplus fixed carbon could be stored. Correspondingly, the amount of CO_2 in the atmosphere would be reduced.

The advent of fusion energy would change energy usage drastically. It would reduce demand for fossil fuels to produce electricity. It would make possible a hydrogen economy that would require no net use of carbon. Some of the fusion energy could be used to capture CO_2 from the atmosphere for injection into geological formations. Alternatively, the energy could be used to convert CO_2 into solid carbon.

When the environment is altered on a global scale, major problems can arise. Careful monitoring and study of the trends in CO_2 is desirable, together with efforts to develop contingency alternatives. The process of providing adequate energy need not lead to catastrophic consequences.

8 July 1983

Acid precipitation is increasingly an important national and international issue with strong financial and emotional overtones. Abundant examples of deleterious effects have been cited, but there is wide disagreement among sincere people as to mechanisms of damage, who is responsible, and how the problem should be ameliorated.

The phenomenon of acid rain is not new. It has been active for more than a billion years. In addition to carbon dioxide, other substances contribute to acidity. About as much sulfur compounds are released worldwide to the atmosphere naturally each year as are put there by humans. In islands thousands of kilometers removed from industrial activity and presumably unaffected by it, rain with pH 4.7 is common. Soils having a pH of 3.5 are formed without human participation in the process.

What is new from a geological standpoint is large-scale burning of fossil fuels. This activity and its effects are concentrated in a relatively small area of the globe. There the anthropogenic contribution of sulfur oxides exceeds that of nature by factors of 10 to 20. Annual precipitation is often equivalent to 20 to 50 kilograms of sulfate per hectare. Nitric oxides play an important role in the conversion of sulfur dioxide into sulfuric acid, and they contribute about a third of the total acidity of the rain.

The most noticeable effect from acid rain is a lowering of pH in thousands of lakes in Scandinavia and eastern North America. Accompanying this have been substantial increases in dissolved aluminum, which is toxic to fish. As a result, some lakes are virtually dead; others are dying. Recently an additional phenomenon has raised great concern. Substantial areas of forest in West Germany are dying. The matter has become a hot political internal issue among the various German states.

In the United States concern is growing about the health, present and future, of forests in the Northeast. Instances of pathology attributable to acid rain have been noted in the

Adirondacks. Were this pathology to become more intense and widespread, the eventual damage would be great.

Some people in agriculture have spoken of acid rain as free fertilizer since it furnishes needed nutrients. Ordinarily, farmers add ground limestone ($CaCO_3$) to their soils. An application of 6 metric tons per hectare will increase the pH of a heavy soil from 3.5 to about 6.5. The 50 kilograms of sulfate per year from acid rain has little effect on a soil after such a treatment. Adding ground limestone to lakes has resulted in restoration of fish populations. The Swedes are now spending $40 million a year for this purpose. In North America some lakes are being treated, but those that are relatively inaccessible are neglected.

Considerable political steam has been building up, particularly in Canada, about acid rain. They export acidic gases to us, but they import far more from us. The imbalance is causing a regrettable bad feeling. In addition, people in the northeast United States take the position that coal-fired utility plants in the Midwest are a principal source of the acid in the rain that has been falling on them. In consequence of these two factors, legislation has been introduced into Congress that would require that emissions from plants in the Midwest be reduced by more than 50 percent. Annual costs for this have been estimated at $5 billion to $8 billion, which would be borne by electricity users. A large number of studies, however, have shown that the Northeast is itself responsible for a large share of its pollution. Indeed, everyone who drives an automobile is a contributor to acid rain.

If long-term damage from acid rain is to be reduced, it will not suffice to use a single scapegoat. Rather, there must be more conservation, better analysis of how to manage, and the development of technologies that effectively reduce emissions while not creating additional environmental problems.

---

14 December 1984

Concern about acid rain continues to mount, ensuring efforts to enact legislation designed to curtail it when Congress meets again. Earlier, impetus for abatement centered around acidification of some lakes and around Canadian protests about transboundary flow of some of our emissions. Lately, attention has been drawn to possible effects of pollutants on forests. Pathology has been noted recently in trees at higher elevations along the mountains from the Northeast to Georgia.

The extent of the problem in the United States is minimal in comparison to that in West Germany. In just a few years observed occurrences of pathology in the forests there have increased from a few percent to more than 50 percent. The news from West Germany is alarming, and its impact is fueling a demand for action here.

One proposal is that SO_2 emissions should be cut in half. The Electric Power Research Institute has estimated that the cost of such a program would be $10 billion a year for many years. In this country, the contribution of SO_2 to the formation of hydrogen ions is about two times that of the nitrogen oxides. If the target of an effort is solely to diminish the acidity of rain, reduction of SO_2 emissions is a logical objective. However, if the goal is to avoid possible pathological effects on forests, the focus of abatement efforts probably should be NO_x. The nitrogen oxides do much more than give rise to nitric acid. They are involved in photochemical reactions that lead to oxidants such as ozone that are highly toxic to plants and trees. Some of these effects have been noted for many years in the vegetation of southern California. In that area, there are negligible amounts of sulfur oxides but substantial quantities of NO_x and photochemical oxidants. Damage from ozone to trees such as ponderosa pine has been extensive and can be seen in trees as far as 120 kilometers east of the urban centers. Studies have shown that chronic exposure to 6 parts per hundred million (pphm) of ozone results in visible damage to some conifers.

Pollution control in West Germany has lagged behind that in the United States. The Germans have no speed limit on their autobahns, and the vehicles do not have catalytic devices to minimize NO_x emissions. Vehicles are by far the greatest source of NO_x, and the total tonnage of NO_2 emitted by all sources is greater than that of SO_2. Measurement of ozone concentration in the Black Forest has yielded a value of about 9 pphm and a peak value as high as 27 pphm. Explanations advanced by German scientists for the pathology noted in their forests include effects of acid on foliage and roots, heavy metals, aluminum toxicity, insects, and diseases. The truth probably lies in a combination of all these, together with photochemical oxidants.

There are many gaps in our knowledge about pollutants. A major one is the relative effects of wet and dry deposition. The amounts of wet deposition are well known. They account, however, for only 20 to 30 percent of total emissions. What happens to the remainder is a mystery.

Another important unknown is the rate at which lakes are being acidified. Little evidence exists that many lakes are undergoing substantial change now. The Environmental Protection Agency is conducting measurements on two thousand lakes to establish a database. The magnitude (large or small) of the acidification problem should be evident in a few years.

Another unknown is the relation of sources to deposition. Many people believe that burning of coal in the Midwest is the source of acidification of lakes in the East. However, local sources are apparently also important contributors to acidification.

These are only a few examples of the lack of knowledge about the effects of pollutants. A determined effort is now being made in the United States and West Germany to better understand the phenomena and to learn how to cope with them. West Germany apparently needs to take quick action, but in the United States there is time to seek knowledge before committing to enormous expenditures that might prove misdirected.

3 June 1983

Humans fear the unknown, when it seems to pose dangers of uncertain dimensions people tend to overreact. Modern intensive communications foster this trend. The so-called hazardous wastes and the attendant publicity about them are an example. Those people who depend on the media for information and guidance have been frightened, some to the point of hysteria. The reality is that the hazard to public health from the waste sites is trivial in comparison to the danger from cigarettes.

That is not to say that toxic wastes do not exist. They are a problem now and will continue to be for a decade or more. Until the public feels that its health is being safeguarded, chemical industry and to a lesser degree chemists will be suspect. Part of the problem is a great increase in sensitivity for detection of contaminants. At one time, parts per million was the limit. Today measurements are in parts per billion, trillion, or even parts per quadrillion. To the average person 300 parts per quadrillion is far more alarming than 0.0000003 part per million.

An essential step in the long-term control of wastes is to minimize their amount and toxicity. For this purpose, major chemical companies have redesigned processes and are capturing and using by-products. Industry has improved methods of waste handling, including incineration and microbiological treatment of dilute waste streams. Better techniques for use of landfill have been developed. Of particular interest are the chemical capabilities of microorganisms. For example, strains have been isolated that can use some of the polychlorinated biphenyls (PCBs) as carbon and energy sources.

The really vexing problems are those of the orphan waste sites. The Environmental Protection Agency has identified about 15,000 of them, of which about 419 have been classified as especially hazardous. The overall task of minimizing the potential dangers from them is difficult and complex. No two sites are alike. In many instances the sites include what an EPA administrator has described as an alphabet soup of substances.

Waste Management

The various components differ by many orders of magnitude in solubility in water, volatility, adsorption by soils, and toxicity. In some instances, the wastes are present in steel drums. In others, they have become disseminated in soils. Some of the sites are located on relatively impervious clay; others are on or close to aquifers. The costs of moving steel drums are small, though few communities are willing to serve as hosts to them. The big costs, as much as a billion dollars per site, would come if large quantities of soil were to be processed. A basic consideration is that a substance must be ingested, inhaled, or absorbed through the skin in order to harm health. The principal hazard is ingestion either through drinking water or in food.

At many sites the hazards could be controlled. The wastes could be covered with an impervious layer and entry from the sides minimized. If there is no inflow of water there will be no outflow. Methods of this kind have been employed at Love Canal. The dump site has been covered by a compacted thick layer of clay. Above the clay is a layer of topsoil with grass. Some water leaks in from the sides, but this will be curtailed when impervious walls are installed. Drainage from the site is passed through activated charcoal and the effluent is clean. The quantities of chemicals collected on the charcoal are small. New York health authorities have continued a follow-up of former residents of the area, including most of those who resided there after 1940. No chemically mediated health effects have been established.

Techniques used in control of leaching from Love Canal are applicable in some but not all of the waste sites. In any event, the chemical and geologic conditions at each site must be evaluated and a plan for meeting the situation engineered before the hazard can be completely controlled. The hazards must and will be minimized, but there will be few quick fixes. Drastic action may be required at a few sites, but more harm is likely to come from hysteria than from chemicals.

SCIENCE AND GOVERNMENT

Between science and government, and between scientists and government, there has been a tentative and sometimes tempestuous relationship. Nowhere is that clearer than in the following selection of editorials, which write a kind of history of the ups and downs, the partnerships and the adversary encounters, that have characterized interaction between postwar science and government in the United States.

In the early 1960s, Philip Abelson pointed out the need for improved scientific advice for Congress—a need that was later met with the establishment of the Office of Technology Assessment. He critiqued the use of scientific advice in the executive branch, endorsing the practices of the National Institutes of Health and the Atomic Energy Commission (an agency that no longer exists) but deploring those of other agencies.

In the era of rapid growth in science budgets, he questioned the tendency of government to favor spectacular projects in place of balanced support and pointed to the distorting effect on university research. He deplored the lack of balance, too, in the points of view of scientists summoned to testify before Congress. He pondered the tendency of government to call on science in time of national peril and then lessen its interest in more happy times.

Later, when budgets for science came under pressure and politicians began to ask science agencies to set priorities for funding, he was critical of the agencies' and the scientific community's failure to do so. He inveighed against anti-industry and anti-government attitudes among academic scientists and urged them to broaden the base of political support for basic research. But he was equally critical of irresponsible behavior in government, challenging Senator Proxmire's attacks on seemingly obscure basic research.

In the 1970s, he criticized shortsighted, on-again, off-again programs for science and technology, noting the difference in political and scientific time scales for effective results in research or in educational policies designed to produce more scientists and engineers. He argued against increasing federal intervention in the universities and in the medical schools. He meditated on the narrow escape from restrictive legislation governing recombinant DNA research—and called on the research community to police itself sufficiently to avoid any incident that could trigger renewed efforts. And he called for scaling down a governmental regulatory apparatus that he described as out of control, citing examples of excess in areas such as exposure to carcinogens of uncertain toxicity.

In the 1980s, Abelson discussed the link between high technology and regional prosperity that has since led to a national scramble to lure high-tech companies to many states.

Throughout more than two decades, Abelson remained a keen observer of the relationship between science and government and a fearless and tart-tongued commentator on folly wherever he found it.

28 June 1963

Increasingly the future shape of science is being determined by legislative actions taken by men and women who can be expected to have only a superficial knowledge of the technical facts embodied in their decisions. The government provides about two-thirds of the support for the nation's efforts in science and technology, and the fraction has been growing.

On the surface it would seem that more money for science is a good thing. Indeed, I have heard some scientists say that it would be almost impossible to provide too much support. However, the realities today are that in many areas of science and technology the crucial bottleneck is brains, not money. For instance, top-quality physicists are in short supply, and this deficiency is likely to continue. When Congress votes to expand activity in a field requiring first-rate physicists, it simultaneously makes the negative decision to remove them from other important endeavors.

It has been pointed out that the educational background of members of Congress is heavily weighted toward the legal profession with little representation from science. The remedy usually proposed is that more scientists should get into politics. On the surface this suggestion is attractive. One obvious difficulty is the time it would take for any considerable group to be elected. A second problem is that, in becoming politicians, the erstwhile scientists would in general lose their professional acuity. Moreover, there is no certainty that an individual trained in science would bring as much wisdom to Congress as one trained in the law. Some of the most narrow-minded, uncompromising, chauvinistic individuals in this world are scientists. Many research workers are deeply convinced that their narrow area of inquiry is the only one worth pursuing. I recently sat on a panel that cheerfully toyed with the desirability of channeling the total gross national product into a single area of scientific endeavor. An individual representative of such a body of opinion would be a dangerous nuisance on the congressional scene.

Devil's Advocates

The government does not suffer from a quantitative lack of scientific information. Rather, the difficulty is that most of the advice comes from special pleaders. The executive branch has good counsel from the Bureau of the Budget and Jerome Wiesner's office, but the Congress has no independent impartial source of advice. Since the legislative branch cannot evaluate technical proposals, the temptation arises to employ phony arguments in advocating major projects. In scientific circles there is a tendency to be more concerned with the glamorous, salable aspects of a proposal than with intrinsic merit.

It seems well to consider other ways of improving the scientific judgments of Congress. To make good decisions it is not necessary to digest all the facts. It is necessary to be well advised. One of the more promising methods would be to make available to Congress a special group of scientific counselors. These would supplement existing staff and would not be permanent government employees. They could be nominated by such a body as the National Academy of Sciences on request of Congress. They might serve for short, intensive periods while retaining their professional connections. They would be expected to act as devil's advocates (the *Washington Star* recently made a similar suggestion), with a duty to insure that the public interests were well protected. If such a system could be properly implemented, a substantial improvement in the quality of science legislation might ensue.

31 May 1963

The federal government is aware of the value of availing itself of the best possible counsel in scientific matters, and most scientists will accept appointments on Washington committees. Unfortunately, most such committees function ineffectively. In part, this is because they are appointed for inappropriate tasks or for inadequately delineated objectives. Thus committees may be asked to ponder the imponderable or to make decisions that timid administrators should have the courage to make. Even when the scope of the committee's functions is proper, a poor outcome may result if the agenda and procedures are badly chosen. Moreover, if a panel produces a wise result, the product is worthless unless it reaches and is acted on by those in authority.

At least two agencies in Washington use scientific advice effectively—the National Institutes of Health and the Atomic Energy Commission. Advisory groups of these agencies have important features in common: long tenure, chairmen not affiliated with the government, preparation of reports by members rather than by the agency secretariat, and free access to agency heads.

Evaluation of the relative merits of applications for grants involving more than a billion dollars a year is made by NIH study sections, which usually consist of about twelve experts who serve for four years and meet three times a year for two or three days. NIH personnel perform executive secretarial service, but the chairmen are outside scientists. Two members look deeply into each application and present their views to the full committee. After discussion, a vote is taken and a numerical priority is assigned by each member. The consensus of the discussion is written up by one or two members responsible for the close study and serves as a permanent record. These outputs have an important bearing on whether a grant is made. Top administrative personnel of NIH appear at the meetings. Morale is high. Members give devoted and thoughtful service and often spend extra hours on their tasks.

Effective Use of Scientific Advice

The General Advisory Committee of the AEC has different and broader functions, but its features are similar, and its activities have led to important advances in the field of atomic energy.

Illustrative of undesirable practices are the procedures of another large agency. This organization has successfully recruited as consultants most of the best talent in relevant fields, yet its committees have little influence on its programs. Appointments are for one year. The committee chairmen and secretaries are government employees. They are conscientious, but their scientific attainments and prestige do not match those of the visiting scientists. Agendas for the meetings are chosen without adequate consultation with members. Often the topics seem trivial in comparison with the topics, not discussed, that need full discussion. The minutes, if any, are fragmentary and are prepared by the secretariat and circulated long after the event. Repeatedly, key ideas are brought forth by consultants, but are lost. Urgent recommendations are made, but seem rarely to be put into effect. Either they do not appear in the minutes, they are not conveyed to those in authority, or they are otherwise neglected. The committee never sees the agency heads and cannot be sure its voice is heard.

In view of these contrasts, differences in the effectiveness and morale of the scientific advisory groups in the various agencies are not surprising. The basic features of the system used by NIH and AEC ought to be adopted by other agencies.

1 February 1963

The legislative process for support of science seems to function best when a spectacular package is involved. Although Congress has attempted to give every encouragement to science over the past decade, there has been particular emphasis on research in medicine, high-energy nuclear physics, and more recently, space. It is almost certain that funds for space research will increase sharply. This is an important frontier, but only one of many. There are negative features of these great spectaculars. The President's request for $98.8 billion is certain to come under attack, but appropriations for defense and space research are unlikely to suffer. Other areas are relatively more vulnerable, and some may receive less money during the next fiscal year. Formerly, when one segment of research was supported on a large scale, other areas also benefited. With research and development appropriations now taking an unprecedented proportion of the national budget, further expansion across the board may not come so easily as in the past.

Another negative feature arises from the fact that the number of competent investigators is limited. The great expansion in space research will in part be accomplished by recruiting workers away from other fields. Many areas of science that have promise of yielding important philosophical and practical results will suffer as talent is withdrawn.

Still another negative feature is a psychological one. Scientists, like other human beings, are affected by fads. They tend to go with the crowd. The research worker who does not go with the crowd encounters a rather bleak climate. He or she is likely to be regarded by administrators and laymen as an odd fellow who is not in tune with the times. Under this pressure, undue emphasis develops on glamorous areas.

Government policies are shaping academic research in this country, but who in government has as his or her primary responsibility the duty to give continuing serious thought to the effects—positive and negative—of excessive concentration on a

Government Support of Research

few areas? Support for research should be balanced and should reflect needs and opportunities throughout science. One organization that could be helpful is only sporadically called on. The National Academy of Sciences is broadly representative of the sciences. Its members are drawn from all sections of the country. Unfortunately, the Academy has recently had little influence in formulating broad policies with respect to science. The organization has been used principally as an agent to generate still more spectaculars such as the International Geophysical Year. The National Academy of Sciences–National Research Council could serve a broader function, and the government would be well advised to avail itself of this source of wisdom and experience.

24 July 1964

In the past year Congress has wrestled with problems of controlling and legislating for research and development funds, but achievements so far appear to be minuscule. One problem that has been scarcely considered is that of obtaining a multiplicity of well-founded opinions on scientific and technical programs. The authorization hearings for the $5-billion space program illustrate the point. The House and Senate committees heard extensive testimony from government witnesses representing, for example, the National Aeronautics and Space Administration. Although a majority of the nation's scientists question facets of the program, no opposing witnesses appeared. In part, this was because they were not invited; in part, because they did not seek a hearing.

In contrast, committees considering legislation in areas other than science and technology often find that many citizens ask to testify. In matters in which conflicts of self-interest exist, issues usually are well examined. For instance, committees considering legislation concerned with labor and management are presented with expert testimony from both sides. Many organizations support staffs that compile reports, prepare presentations and rebuttals, and look for special opportunities to advance their cause. In an effort to triumph, the opponents may propose fruitful new ideas. The public and Congress emerge with a sounder view of the factors involved.

There are several reasons why scientists do not seek to testify at hearings on research and development legislation. Most scientists are unaware of the schedule of impending hearings and unfamiliar with mechanisms for obtaining an audience. Only a limited number feel competent to make judgments in the diverse, highly technical areas. With few exceptions there are no staffs to aid in preparation of material. Only when highly emotional issues, such as fallout, are involved is there a semblance of broad response. A major reason why research and development legislation is not more adequately discussed is a lack of evident immediate clash of self-interest among scientists.

Only One Side of the Question

The self-interest of those who advocate expenditures is obvious, but who makes the probing counterargument? At present, it comes not from informed witnesses but from members of Congress, and their principal well-worn line is that we are spending too much money on science in general.

Failure of scientists publicly to criticize to any appreciable degree programs many deem ill-judged often stems from analysis of the balance sheet of their own self-interest. On the positive side is the consideration that the long-term interest of their profession and the nation dictates that unwise expenditures not be made. If the public loses confidence in the integrity of scientists, the sequel could be calamitous for all. But this is a nebulous possibility that does not outweigh the realities of the present. The witness in questioning the wisdom of the establishment pays a price and incurs hazards. He or she is diverted from professional activities and stirs the enmity of powerful foes. Reprisals may extend beyond the witness to his or her institution. Perhaps the witness fears shadows, but in a day when almost all research institutions are highly dependent on federal funds, prudence seems to dictate silence.

16 October 1964

In this October of 1964, the Cuban crisis of two years ago seems almost like a bad dream. The Nuclear Test Ban Treaty of a year ago has been quite effective in easing world tensions. While vestiges of apprehension remain, billions of people lead freer, happier lives in the new political climate. If this relaxation continues, a variety of adjustments will surely follow. For instance, the urgency for defense spending will diminish. At the same time, there will be a change in the public's attitude toward science.

For nearly twenty years, most people have identified science closely with defense. To a large degree these public attitudes were conditioned by wartime technological developments and especially by the atomic bomb. A measure of the extent of the people's faith in science as a shield was the reaction to *Sputnik*—not only at the time, but also subsequently. One indication was the suddenly enhanced position of scientists in the high councils of government; for example, elevation of the status of the President's Science Advisory Committee. A second and perhaps more significant measure was the grass-roots response with respect to education. This has produced changes in school curriculums and course content that may be the most important longtime residue of the event.

With imminent mortal peril receding from view, how will the public regard science? In a large measure this will depend on the values which, to scientists themselves, characterize their work.

There are practical reasons why science should have an honored place in society. The base of our present prosperity rests on science and associated technology. While our attention has been over-focused on military problems, new competition and new rivals have appeared. In international trade the Germans and the Japanese are increasingly successful as they skillfully exploit applied science. In this arena they leave the Russians far behind. I shall never forget the half-frightening impact of an international industrial fair at Frankfurt and conversations there with Americans who were in competition with West Europeans.

An especially important area where science will continue to be the servant of progress is medicine. All mankind faces deadly enemies in the disease processes that have killed and maimed far more individuals than have died in wars. While there have not been recent major dramatic practical consequences of medical research, our improving basic understanding of biological processes guarantees future benefits for all. At the same time, multiple small victories have lessened the menace of even such formidable enemies as cancer and cardiovascular disease.

The most neglected but perhaps the most important reason for carrying on scientific work lies in the spiritual sphere. Today most of the earthly land frontiers have been explored. Where can society look for innovation? For new challenges? We must change continuously to some degree, or we shall stagnate. One of our best sources of innovation is science and technology, for the spirit of science is innovation. One other value of science has not been discussed much of late. It stems from man's hunger to know. Even today we are faced with many unknowns. These represent an invigorating incentive to man and to science. When science drives back the frontiers of ignorance, it does so for all, and all can enjoy a greater sense of human dignity for it.

If we are to move forward toward a great society, we must have great challenges, great thinkers, and great innovators. Science can furnish its share of them.

2 July 1965

Distribution of research and development funds is becoming a major political issue. One reason is that present distribution is very uneven. In 1963 one-half of the fifty states received 96.8 percent of all federal R&D funds while the other twenty-five divided the remaining 3.2 percent. The fifty senators from the have-not states may be expected to attempt to improve their states' share. This tendency will be reinforced by a growing general awareness of the role of science and technology in our changing civilization.

The extent of the beneficial effects of federal funds is not clear, though in some instances the economic stimulus has seemed disproportionately large. In other instances benefits were nominal. However, many of the states whose economies have grown rapidly are recipients of large sums of federal R&D funds. In contrast, some regions of this country not receiving large R&D contracts seem on the road to becoming new Appalachias.

In a recent speech at a Midwest Governors' conference, Charles Kimball, president of the Midwest Research Institute, outlined problems of one such area. The Institute had made a study of a six-state region, including Arkansas, Iowa, Kansas, Missouri, Nebraska, and Oklahoma. During the decade 1950–1960 the number of jobs in the area had increased only 2.3 percent as against 14.5 percent for the nation. The average family income is now much below the national figure. At the same time more than a million people moved to other parts of the country. Although total population increased slightly, the increase was much less than the national average.

Contributing to discontent over uneven distribution of R&D funds is growing frustration over a Middle West "brain drain." For instance, in Iowa the two major universities estimate that from 55 to 75 percent of the graduates leave the state for their first jobs. The states in which the "Big Ten" universities are located produce about 40 percent of all scientists and engineers with advanced degrees, but this region receives only a small fraction of federal R&D funds. To apply their training many of the scientists must migrate; their migration represents an economic loss to a region that yearly spends huge sums on higher education.

Some of the frustration felt in the Midwest was expressed recently by Fred Harvey Harrington, president of the University of Wisconsin, when he said, "failure to provide for geographical spread of federal defense and space contracts has brought our nation face-to-face with a most serious kind of over-concentration." Two examples he cited were "over-concentration of our working scientists and engineers by region" and "over-concentration of our industrial and defense strengths by region." He went on to say, "if these tendencies continue in the award of federal research, development, and production contracts, we are on the road to manpower chaos, economic chaos, social chaos, defense chaos."

The feelings expressed by President Harrington are not unrelated to pressures now evident with respect to the proposed new 200-Bev accelerator. Major midwestern universities have agreed to support whatever midwestern site appears to be most in the running after initial screenings. If the region exerts its maximum potential pressure, Washington will find it difficult to place the facility elsewhere. Whatever the outcome, politics is destined to be a crucial factor, and an important precedent will be established. The influence of political and geographical considerations on the allocation of research funds is certain to increase.

28 January 1966

Federally supported research activities are being reexamined. The immediate cause is the budgetary squeeze brought on by the war in Vietnam. More fundamental is the fact that a twenty-year honeymoon for science is drawing to a close. Although needs for support of basic research are increasing, expanded budgets will be obtained only after convincing justification has been provided. Indications of the present climate can be seen in the executive branch, the press, and Congress. In contrast to other years, President Johnson hardly mentioned science in his State of the Union speech. In an article in the December issue of *Discovery*, John Finney, influential Washington correspondent of the *New York Times*, was critical of science policy-making. Two congressional committees have expressed their concern. In hearings held 7, 10, and 11 January 1966, a subcommittee headed by Representative Henry S. Reuss asked: "Are we matching research and development programs with national goals?"

The most significant development affecting science is the content of the report of the Daddario subcommittee. This document is the result of a searching examination of the National Science Foundation. It is especially critical of the quality of national science policy-making. It urges a new role for the National Science Board, and it suggests that NSF should become more interested in applied research.

Two years ago, the Daddario subcommittee sought answers to the following questions: What level of federal support is needed to maintain for the United States a position of leadership through basic research in the advancement of science and technology and their economic, cultural, and military applications? What judgment can be reached on the balance of support now being given by the federal government to various fields of scientific endeavor, and on adjustments that should be considered, either within existing levels of overall support or under conditions of increased or decreased overall support?

In one form or another these questions have been posed by Congress since 1950. Originally NSF was expected to provide the answers, but it managed to avoid the problem. Congress did not press the issue, for the funds devoted to science were relatively small, and the honeymoon was on. Apparently despairing of getting a response from NSF, President Kennedy in 1962 assigned the problem to the President's Science Advisory Committee, Office of Science and Technology. When this group was not sufficiently responsive, the Daddario subcommittee put the questions to the National Academy of Sciences. The Academy recommended a 15 percent annual increase for support of basic research and suggested use of NSF as a "balance wheel," but did not provide mechanisms for allocating funds among the various branches of science.

The new Daddario report implicitly is critical of this failure. How can NSF act as a balance wheel if no one knows what constitutes balance? Having failed to obtain what it considered satisfactory guidance from NAS, the Daddario subcommittee has now turned to another source. The NSF Board has been selected as the new fount of wisdom or, perhaps more accurately, the holder of the buck. The report, however, calls for a diminished role of the Board in the management of NSF.

With its comments and its recommendations of drastic changes the report conveys congressional impatience with key elements of our scientific leadership. As the government agency charged with fostering basic research, NSF has a special responsibility to lead in formulating and illuminating science policy. It must also be a more skillful advocate of the benefits that support of basic research yields the nation.

1 July 1966

For two decades basic research has been living largely on society's goodwill; there have been no major miracles. Although research has made significant advances that in sum have more than justified its support, few of its spokesmen have bothered to do a good job of showing that basic research is currently paying off. Results of this lack of diligence are now evident.

There have been significant changes in the government's attitude toward basic research. President Johnson has called on the National Institutes of Health to plan for "specific results in the decline in death and disabilities" from cancer and heart and other diseases. Much basic research has been done in these areas, the President said, but the "time has now come to zero in on the targets." Congressman Emilio Q. Daddario is pushing the National Science Foundation toward applied work. Key spokesmen of other major agencies, such as the Department of Commerce and the Defense Department, have called for greater emphasis on applied work and, by implication, less on basic research.

Two current factors could place added pressure on basic research. One is Medicare, and the other is a shortage of personnel for applied research. Washington fears that there may not be enough doctors available when the new law goes into effect on 1 July. Why not cut back on medical research to meet the crisis? This would make good political eyewash, although it would add barely 1 percent to the nation's supply of practicing physicians. There is an acute shortage of physical scientists to fill jobs in industry. Why not cut funds for support of basic research by the National Science Foundation? Such a move might increase the applied research manpower pool by as much as 1 percent.

At a time when those who understand the value of basic research should be united, such unity does not exist. Outside the university one finds considerable antipathy toward the academic establishment. Within it, professors have looked down on nonuniversity research, have regarded its practitioners as inferiors, and have attempted to curtail their activities. Most university science graduates must eventually find employment in nonacademic posts. When they do they accept for themselves what they have been taught is a second-class status. As a result they can have deep loyalty neither to their alma mater nor to their employer.

These campus attitudes are unrealistic and destructive. Important research is being done in industry, in government laboratories, and elsewhere. In many areas of physical science, work at industrial laboratories is unsurpassed. In many aspects of biomedical investigation, work at the National Institutes of Health is in the forefront. Similar statements could be made about other governmental and nonprofit research establishments and the national laboratories.

In the present situation major blunders could be made, weakening the entire fabric of science, medicine, and technology. In downgrading basic research, the government could repeat the unhappy experience of the petroleum industry. In 1958 many geologists were dismissed in an economy move. In the next few years, enrollment in geology departments dropped to a small fraction of its former level. Today, the industry wishes to employ far more graduates than are available or will be forthcoming in the next several years.

Attitudes toward basic research are in transition. Industry, currently aloof, could find its vital interests severely damaged while it sat watching. The academic community has some fence-mending to do and should get about doing it.

9 June 1967

Members of Congress serving on committees dealing with aspects of research and development generally become knowledgeable about their areas of responsibility. As a result, they are often well disposed toward support of scientific research. However, as politicians they cannot afford to be so partisan as to become vulnerable. They must take into account tides of public opinion and matters likely to affect opinion, such as articles in mass-circulation magazines.

Several members of Congress have commented privately on the adverse impact on their constituents of an article entitled "The Great Research Boondoggle," that appeared recently in a monthly publication. As a partisan document, the article is a triumph. Research is confused with development, and the reader is left with the impression that the annual cost of government-supported research is $16 billion. Then research is downgraded by citation of examples likely to seem ridiculous to the reader and by skillful choice of guilt-connoting words—such phrases as "federal research craze," "complex jungle of federal research," "sprawling research program," "research bug," "lucrative contracts," "profitable parasite industry," and "getting fat at the public trough."

In the article twelve specific government-supported projects are cited as examples. Most of these involve the social sciences, which receive a tiny fraction of the funds. For example, the article quotes Senator William Proxmire as attacking the National Institutes of Health as a "worst offender" for supporting projects designated "A Social History of French Medicine, 1789–1815" ($11,782); "Emergence of Political Leadership: Indians in Fiji" ($10,917); and "Changing Patterns of [Moslem] Family Life" ($28,755).

Similarly, the Department of Agriculture was cited for spending five years "revising pickle standards." The Office of Education drew mention in an unfavorable context because it

A Partisan Attack on Research

supported research on "understanding the fourth-grade slump in creative thinking."

A knowledgeable observer might smile at so much ado about so little. However, a less astute reader could be left with the impression that a large fraction of NIH and other federal funds is spent in irrelevant areas.

No enterprise supported by the federal government should be free of criticism. Research is no exception. Some scientists have questioned aspects of the science establishment with the goal of making government expenditures more effective. It is desirable that such self-policing continue. However, the article in question illustrates a cost of public self-criticism. It seems very damaging when it quotes a prominent scientist out of context as saying that federal support has encouraged "shoddy, redundant, uncritical and ill-conceived research." The article is also very damaging when it quotes a professor of chemistry at a large university as saying that government support of research is "potentially the most powerful destructive force the higher educational system has ever faced."

Members of Congress can easily judge the validity of magazine articles. Politicians, however, cannot be expected to assume all the burden of setting the record straight. Scientists must help ensure that the public has an accurate understanding of what it is getting for its money.

26 January 1973

Harvey Brooks and others have commented on the mismatch of time constants of technology and politics. A substantial innovation usually requires eight to ten years to reach fruition. Politics has a large emotional content whose thrust changes rapidly and unpredictably. In the course of a decade we experience scores of major or minor practical tempests.

Because of the performance record of science and technology, politicians are inclined to call on them when political problems emerge. Often, though, before substantive efforts can be made, the political climate has changed and a program that was a political asset has become a liability.

The vagaries of the interaction of politics with technology are illustrated by a series of events that began in 1971 and are still in process. In 1971, several factors combined to create a climate in which it seemed politically desirable for the government to foster new technological initiatives. A deteriorating balance of payment carried with it the implication that our technological supremacy had slipped. Widespread and publicized reports told of unemployment among scientists and engineers. There was a general feeling that some of the technological expertise that put men on the moon should be devoted to solving urgent domestic problems. The economy was in a slump, and means for stimulating it were being sought.

Task Forces were formed, and suddenly Washington watchers were aware of the name of Magruder. There was great moving in and out of Washington of distinguished scientists and engineers and talk of programs costing billions of dollars. In November and December 1971, excitement reached a peak and we were told that big things would be announced early in the following year.

Somehow the promised events did not quite come off. The fiscal year 1973 budget request and a subsequent special presidential message on science and technology contained references to technological initiatives, but the presidential requests did not

Technological Initiatives and Political Realities

match the rumors. It is not easy to delineate the programs that resulted from the 1971 excitement. One new budgetary request that was enacted was a $44-million Experimental Technology Incentives Program. Under this program, the National Science Foundation (NSF) and the National Bureau of Standards (NBS) were authorized to develop experimental contract programs to study means by which the federal government could best stimulate research and development. In the words of Lewis Branscomb, this was to be an "opportunity to evolve and demonstrate an economically effective and politically acceptable relationship between federally sponsored R&D and commercial business." It is possible that this program was not the best way to spend federal funds in behalf of R&D. However, the matter will probably not be brought to a full test, for 1973 has brought new political realities. The economy is more robust. Talk of unemployment of scientists and engineers has abated. The President has won reelection. The big push in Washington now is to hold federal expenditures to $250 billion this fiscal year. Congress appropriated about $260 billion, of which about $175 billion is nondiscretionary—for example, interest on the debt. Thus if $10 billion is to be cut, it must come from the $85 billion in which are included the expenditures of NSF, NBS, and other science-oriented agencies. In consequence, funds earmarked for the Experimental Technology Incentives Program have been withheld by the Office of Management and Budget (an arm of the President). The whole affair reminds me of a rhyme I heard as a boy. The King of France and twice ten thousand men marched up the hill and then marched back down again.

6 April 1973

The phasing out of the training grants and fellowship program at the National Institutes of Health will sharply curtail the government's direct support of predoctoral and postdoctoral education in the sciences. Some training will continue in connection with research grants, but the number of students supported will be a minor fraction of those who had received stipends earlier. The dismantling of the government's fellowship program liquidates some excesses but, on balance, is a destructive move, and it comes at a time when the need for some kinds of scientists and engineers is actually growing.

After *Sputnik* was launched, this nation engaged in a frantic effort to expand its scientific capabilities. For a number of years government funds available for research in the physical and biomedical sciences increased rapidly. At the same time, the Apollo program was implemented. These developments created a shortage of scientists and engineers. The government responded by initiating and expanding support of many kinds of fellowships and training programs. To meet the opportunities of the times, universities expanded their faculties, thus increasing opportunities for employment. Industry found it difficult to attract qualified personnel. Demand for scientists seemed insatiable. Help-wanted ads in the *New York Times* and other publications reached record numbers. An index of employment opportunities, based on such evidence, peaked in 1966 at 190 percent of 1961 levels. When government support ceased to grow, demand for scientists began to drop. Universities no longer needed to expand their faculties, industry began an era of retrenchment. The Apollo program entered its final phases. Suddenly there were unemployed scientists and engineers, and the index of employment opportunities dropped below 40 in 1971. During the peak years, it was common for top-quality graduates and Ph.D.s to receive dozens of job offers. In 1971,

Blow-Hot, Blow-Cold Educational Policies

the best students often had only two or three opportunities, and some graduates had no jobs for months.

The most dramatic unemployment problem was in the aerospace industry. When activities were cut back, severe local unemployment resulted. A picture of an engineer driving a taxi created a profound and lingering impression.

Those in government who wished to dismantle the fellowship programs had a useful excuse. Why train scientists when there were scientists unemployed? To a substantial extent, the unemployment argument is no longer valid. The index of employment opportunities has climbed above 100. In some regions there already are shortages of engineers.

Influenced by current antitechnology talk and by reports of unemployment, first-year college students have been shunning engineering. Beginning enrollment is down more than 30 percent from two years ago. Students often leave engineering courses; they rarely enter them after the freshman year. Thus, a severe shortage of young engineers may now be projected four years hence. This is likely to come at a time when this nation will be engaged in frantic "crash" programs to solve the energy crisis—an effort that will involve a tremendous construction program and large numbers of engineers.

In the years ahead, this nation will encounter many unexpected problems requiring the skills of scientists and engineers. We may well come to regret bitterly the fact that we have been unable to do better than follow destructive blow-hot, blow-cold educational policies. We should adopt the more realistic assumption that this nation must have good science, good medicine, and good engineering, and we should make it possible for the top students, regardless of financial ability, to participate.

---

17 October 1975

University presidents and other spokesmen are beginning to state publicly what they have been saying privately. Congress and the federal bureaucracy are increasing their many modes of interference with universities. No institution is immune, and indeed the more prestigious one is, the more it is an object for attack. A common device is the ultimatum with a short deadline: If you do not do such and such, your grants and contracts will be cut off.

For some schools the confrontation is not dramatic, it is piecemeal. There are at least twelve federally mandated programs that cumulatively impinge on the financial health of all universities. The American Council on Education has stated that one large public university's annual expenses for implementing federal programs increased from $438,000 to $1.3 million between 1965 and 1975. During the same period a large private university's expenses increased from $110,000 to $3.6 million, and a private college's $2,000 to $300,000. The monetary expenditures are only part of the costs. They do not reflect the diversion of effort from scholarship to attention to federal demands.

Until about 1960 government involvement in academia was not great and interference was minimal. But in the late fifties, federal grants for research started to become a substantial factor in university budgets. The government chose to demand detailed accounting for individual grants. Since that time, the fastest growing component at many universities has been the business office. The sixties also brought a weakening of the status of presidents of universities. A contributing factor was the Vietnam War, but the federal grants system also played a major role in diminishing the authority of university leaders. In addition, the sudden termination of large federal fellowship programs that had previously grown rapidly caused substantial financial problems.

Thus, in the seventies the leaders of universities were ill-equipped to deal decisively with Washington and its agents. In

Federal Intervention in Universities

consequence, the universities are now forced to cope with laws, proposed laws, regulations, proposed regulations, and authority-grabbing bureaucrats. The laws are proposed and enacted for worthy purposes, such as occupational safety, fair employment, or social security. Each of itself is laudable and defensible. But their total impact on the financial and intellectual life of the universities is severe. Moreover, the laws are subject to interpretation by the executive branch. Enforcement of regulations is in the hands of local agents, who often extend federal interference with university affairs. For example, auditors from the San Francisco office of the Department of Health, Education, and Welfare have been pushing around the California State University and College System. They demand that anyone paid on a federal project account for his or her total effort and that the schools change their payroll systems, under the threat that noncompliance will result in withholding of letters of credit.

A saddening development in the federal approach to universities in the past decade has been a shift from offering inducements to threatening punishments. This is especially significant in the area of fair employment practices. The universities have been slow in recruiting women and minorities, but bludgeoning and threats are creating a poor climate for change. Competent women appointees are being taunted that they owe their positions not to their own qualifications but to federal pressure. How much better change might have gone with the carrot instead of the stick!

The irony of punitive federal intervention is that a government that is unable to manage its own affairs competently insists on spreading its own brand of inefficiency throughout higher education. It is to be hoped that the university faculties will unite behind their presidents in opposing further federal involvement. A truly unified academic community could halt the federal crippling of higher education.

16 September 1977

When federal support for academic research and education was proposed, there were many misgivings about eventual government intervention. This was slow in coming, but lately it has been highly manifest. As a result, Washington has become to many an object of fear and antipathy. This is especially true of the deans of medical schools. Interference has reached such proportions that some institutions are now willing to forgo federal funds.

The medical schools have crucial roles in research, teaching, and health care. They are leaders in applying biomedical research. They teach the latest and best material to the students. Patient care at their hospitals sets standards for excellence. Among medical practitioners, those associated with medical schools have been tops in their fields. The medical schools have been responsive to societal desires. In an ideal world such citadels of virtue would be trouble-free. But this is not an ideal world. Too often, excellence is a magnet for trouble. In the case of medical schools, most of their problems have come from trying to do too much for society. Resultant financial strains have made them vulnerable.

Research activities are a drain, not a bonanza. Tuition covers a small fraction of the cost of medical education. General practitioners and others refer their costly or botched-up cases to the medical schools.

A major source of deficits has been in the educational activities. A decade ago there was much talk about a shortage of doctors. The medical schools responded positively and set about increasing their enrollments. This involved capital expenditures only in part made good by the government. It also entailed expanded faculties. In response to societal wishes, the schools held down tuition fees so that worthy but less affluent students might be served. While costs per student were in the range $10,000 to $20,000 per year, the median tuition was about $4000. Thus, federal subsidies for tuition (capitation) were eagerly sought and accepted. At first these were fairly liberal, but in the academic year 1976-1977 they amounted to only $1000 per student.

Congress has chosen to use this pittance as leverage in an attempt to control admissions policies at medical schools. Under current legislation, each school must admit an increased number of third-year students (10 percent of the class or ten, whichever is greater) to obtain a capitation of $2000. The increased capitation does not cover the increased expense. The bulk of transferees would come from foreign medical schools, where standards of admission and training are generally inferior. Most of these students are U.S. citizens who were initially rejected by our medical schools. However, many are offspring of wealthy parents who could afford to send their children abroad for training in the expectation of later returning to practice in the United States. The medical schools should reject the new strings on capitation as a matter of principle as well as on financial and egalitarian grounds.

There is a broader, important issue. That is the long-term costs of increasing further what is now regarded as an excessive number of doctors. Given a large body of hypochondriacs and lonely people, and given third-party payments, there is practically an infinite demand for medical attention. Eli Ginzberg has cited estimates of the total expenditure society makes in supporting a physician for a year ($250,000). Thus, during one professional career, society will, on average, spend about 8 million current dollars. For each year that the present capitation legislation is implemented, the cost to society for the rejectees will be in excess of $10 billion.

It is to be hoped that the medical schools will be steadfast in their refusal to accept further government coercion.

13 January 1978

During 1977 the scientific community escaped a threat to the freedom of inquiry in the form of harsh legislation. The ostensible target was alleged hazards of recombinant DNA, but objectives of some of the proponents were broader. The escape from restrictive legislation may prove to be only temporary. Last year congressional action was delayed in part as a result of extremely effective lobbying by scientists, especially a group headed by Harlyn O. Halvorson. If biologists relax, the battle could be lost. Moreover, irresponsible acts by individual scientists could be very damaging.

One of the ironies of the situation is that biologists drew lightning to themselves. As long ago as the early 1960s some leading biologists were warning of ethical problems they envisioned as arising from genetic engineering. These warnings proved premature, but they were given prominence in the media. Statements discounting the imminence of genetic engineering received little attention. Gradually the public became uneasy about a hazard it could neither evaluate nor, perhaps, control.

The recombinant DNA technique that became available in 1973 opened new vistas in genetic research. It made possible the preparation of large amounts of individual genes. It also made possible the incorporation into the genome of chemically synthesized pieces of DNA. Molecular biologists who first became aware of the new developments could envision all kinds of experiments, some of which they felt might produce new pathogens. Seeking to be responsible citizens, they called attention to the matter and recommended a moratorium on some experiments.

In July 1976 the National Institutes of Health published guidelines that were soon made applicable to all research performed under federal grants. The guidelines permitted use of certain nonpathogenic mutants of the K-12 strain of *Escherichia coli* for recombinant DNA experiments. The containment procedures required were reasonable and adequate.

However, the long series of warnings about genetic engi-

Recombinant DNA Legislation

neering had created a climate of public opinion favorable for critics of recombinant DNA research. Though relatively few in number, their influence was great. The relevant committees of Congress accordingly prepared restrictive legislation. Because of the pressure of other business, Congress did not act quickly. In consequence, there was time for lobbying against the bills. In addition, during 1977, Roy Curtiss III produced further information that minimized potential hazards arising from the K-12 *E. coli* mutants. Halvorson and others pointed to the fact that extensive work with pathogens at Fort Detrick and the Centers for Disease Control in Atlanta had not led to contagion among the families of microbiologists. Stanley Cohen showed that nature was already performing many of the experiments that the legislation proposed to regulate.

But some kind of legislation seems likely. At present, industrial laboratories are not compelled to follow the NIH guidelines. However, in the process of regulating industrial laboratories, almost anything can happen depending on the public mood of the moment. When such legislation is finally adjusted in a conference committee of the House and Senate, strange provisions can enter that bear little relation to the original bills.

A major hazard is that during the crucial moments of the legislation, news will come out of some irresponsible act by a scientist engaged in recombinant DNA research. This need not be an act of substance. Already at the Stevenson hearings in November, it was made clear that failure to complete some paper work could draw censure.

Today recombinant DNA research is highly productive, highly competitive. Workers are under temptation to take shortcuts. But they should behave as if their every act is under scrutiny, for indeed it is—by assistants, colleagues, or competitors. A scientist who furnished the pretext for restrictive legislation could count on the ill will of many of those he or she most wants to impress.

---

5 May 1978

Detailed government control of many aspects of society continues to expand. It is on an exponential growth curve that must be abandoned if further decay of the health of the economy is to be avoided. At least part of this country's loss of competitive ability in world trade must be attributed to the drain of the tens of billions of dollars wasted on excessive mandated paper work. Probably more lethal and harder to combat are inefficiencies and delays stemming from government regulatory agencies.

Growth of the government's role is indicated in many ways. One is the greatly increased number of employees; another is the expansion in funds spent by federal, state, and local governments. In 1929, this amounted to 9 percent of the gross national income. By 1960, the fraction was about 17 percent, and by 1976, 28 percent. In monetary terms the contrast is even more striking. Since 1929 there has been about a 125-fold increase in the federal budget. Further indications of government trends are the number of pages in the *Congressional Record*, the total number of words in legislation enacted by Congress, the pages devoted to the regulations printed in the *Federal Register*, and the number of regulatory agencies. And the tendency to expand is accelerating.

The trend toward increased government complexity has not gone unnoticed and indeed seems to be resented by a substantial fraction of the public. An earlier belief that the federal government could cure any social ill has been replaced by the view of many that most federal programs are relatively ineffective and consist mainly in "throwing money at problems."

Nevertheless, the congressional mill grinds on, turning out more complex, ill-fated legislation. When bills are drafted, the objectives are set forth in eloquent, unassailable phrases. The nitty-gritty of the matter, though, is something else. It is detailed, complex, fuzzy, and usually contains provisions that only a Solomon could implement.

What Congress seems to disregard are the limitations of the people who will implement the legislation. Mere mortals must interpret the vague language of bills and translate it into regulations that must be administered. Today, as many as one hundred thousand federal employees are engaged in regulatory activities. Few are geniuses, few are saints. Like many other humans, they seek to excel, they wish to be important, they wish to extend their influence and authority. When dealing with those outside the government, they don the robes of federal power. All too often a person of modest attainments is in a position to make decisions involving hundreds of millions of dollars. In this situation there are basically three choices. One can say yes, by which action one abdicates power. One can say no, which might be appealed and lead to a stink. The safe course is to ask for more information. If sufficiently diligent in this respect, the regulator will be overwhelmed with hundreds of pages to read and will obviously require assistance to perform the chores. Thus, the way can be prepared for a promotion and higher pay.

We have created a regulatory machine that is unmanageable by the President and his cabinet officers. The situation today is much worse than that which on one occasion faced President Johnson. There was a leak of information that annoyed and concerned him. In vain, he ordered a great effort made to identify the culprit. To a friendly visitor he later exclaimed, "If I could find the son of a bitch I'd fire him."

Congress had the wisdom to create an Office of Technology Assessment, designed to attempt to foresee adverse and beneficial consequences of new technology. Congress should take a leaf from this book and make searching studies of the consequences, both retrospective and prospective, of its actions. No major piece of legislation should be enacted without detailed realistic consideration of the mechanism of implementation and a study of its direct and indirect impact on the economy. Were Congress to do this, it might often find that drastic changes in the terms of its bills were indicated.

13 October 1978

A small but increasing number of academic scientists have become alarmed at the threat of crippling federal regulation of their laboratories. They are concerned about rules in the process of being adopted by the Occupational Safety and Health Administration (OSHA) governing exposures to suspected carcinogens. The rules will be applicable to all industrial laboratories and most colleges and universities. No one quarrels with the desirability of guarding against exposure to chemicals of known toxicity. Controversy arises when substances of limited toxicity are placed on OSHA lists and users are to be subject to regulations designed for very dangerous substances.

In compiling its lists OSHA spurned repeated offers of help from the National Academy of Sciences. Instead it employed a private firm to make a literature survey. In reviewing each chemical, if the firm found two positive results for carcinogenic or neoplastic effects, it looked no further. It stated that negative findings were not taken into account.

At the moment, OSHA asserts that its lists are tentative. However, once a substance is placed in category I, it can be removed only by the Secretary of Labor, who in turn must follow stringent guidelines.

Items are assigned to category I when they produce cancer in humans or in at least two species of test animals. Hematite (iron oxide) was given this classification on questionable evidence. Other substances in the group include benzene, chloroform, carbon tetrachloride, and a host of nonvolatile inorganic salts.

Items are placed in category II if there is suggestive evidence for production of cancer. Ethyl alcohol is one of the items in this category. Cases of liver cancer have been noted among alcoholics. However, the concentration ordinarily present in a laboratory atmosphere is less than that at a cocktail party. A practical regulation for volatile substances such as benzene would be to require that operations with them be conducted

Regulating Exposure to Carcinogens

in an efficient hood. This would eliminate any hazard.

The proposed OSHA rules for handling substances in categories I and II are set forth in the *Federal Register*, written in gobbledygook apparently by lawyers for lawyers. From a large number of pages the following was extracted by William P. Schaefer of the California Institute of Technology:

> The employer must monitor each workplace to determine the concentration of the toxic substances at least quarterly. Each employee must be given in writing a report of his or her exposure.
>
> The employer must provide at no cost to the employee, and ensure that employees wear, protective clothing and other equipment. The clothing must be kept clean by the employer, and the employer must ensure that the employee removes all protective clothing at the end of the work shift.
>
> The employer must institute a program of medical surveillance, including individual medical examinations and tests for each employee exposed to any toxic substance, all at no cost to the employee. There must be an initial examination and periodic examinations thereafter.
>
> Records must be kept of all monitoring in the workplace and of all employees who are exposed to any toxic substance. The records must be kept for forty years, or twenty years after termination, whichever is longer.

One of the galling features of the OSHA regulations is that professional scientists who have only occasional exposure to chemicals such as benzene must comply with rules designed for untrained workers exposed chronically. Were the regulations designed realistically, the several hundred scientists working with dangerous carcinogens would be carefully protected while expensive and useless annoyance would be avoided for perhaps one hundred thousand others whose most serious laboratory exposure is to ethyl alcohol. If OSHA had its eyes on the main chance, its highest priority would be to ban cigarette smoking in all workplaces. The present OSHA proposals invite ridicule, contempt, and noncompliance.

---

21 January 1983

Unemployment was a dominant issue in many states in the last election, and it could remain so for some years. The older industries, such as auto and steel, may eventually partially recover, but they face great international competition. Governors of states are under pressure to do something that promises to lead to more jobs. Many governors are pinning their hopes on high-technology industries, which have grown while other industries have been stagnant or decaying. The National Governors' Association has sponsored meetings and committee work on the topic. About half of the governors are fostering some kind of activity, such as the formation of an advisory council on high technology, in their own states.

Representing the National Governors' Association, Charles S. Robb of Virginia stated last year in testimony before a congressional subcommittee that "the industrialized world stands on the threshold of a technological revolution that will change the American way of life and the composition of the nation's work force as much as the industrial revolution did a century ago. . . . Our ability to lead this technological revolution, as indeed the United States led the industrial revolution a century or so ago, will bear directly on our share of world markets—a share that will continue to erode unless we act promptly and wisely." Governor Robb also touched on the importance of interactions between universities and industries in fostering innovation in high technology.

At a juncture at which governors are under pressure to increase jobs, they find themselves with limited resources. At the same time, outlays for education are large. They are aware of activities around Route 128 in Massachusetts and near Palo Alto in California. They have to ask themselves whether their state universities can do what Stanford and the Massachusetts Institute of Technology have done for their regions. If the recession continues, other universities can expect increasing

High-Technology Jobs

pressure and questions from governors and legislators.

There is a large gap between a belated recognition of the importance of high technology and achieving something in the way of jobs. The translation of research into substantive applications usually takes a decade or more. The transformation of small innovative companies into giants takes time. Governors may be well advised and have great plans, but their tenure is limited. Many were swept out of office in the last election. Their successors will wish to formulate their own programs.

For alert states there may be a partial solution for some economic problems. Many of the high-technology companies currently centered on Route 128 or in Silicon Valley are looking elsewhere for expansion as costs of labor, housing, and land have become excessive. A congressional staff study describes responses of 671 companies to a questionnaire concerning factors that influence their decisions to locate facilities.

The high-technology companies are science-based. Research and development outputs are more important to them than to other manufacturing industries. Major determinants in their decisions to locate facilities include availability of skilled labor, labor costs, and state and local taxes. Other factors include community attitudes, costs of property and construction, transportation systems, available area of expansion, good schools, and proximity to recreational and cultural resources. The study indicates that high-technology companies plan to expand at highest rates in the Midwest, Southeast, Southwest, and Mountain and Plains states. Where they will actually locate may well depend on local initiatives. Michigan, North Carolina, and Arizona have been especially active in seeking to foster high technology and are meeting with some success. In the majority of states there has been more talk than action.

INSTITUTIONS

From *Science* itself to the universities, the industrial research laboratories, the libraries, the instrument makers, even the science press, Philip Abelson has cared about science and its institutions—how and where research is done and by whom, how it is reported in technical journals and in the media, how it is perceived by the public.

In the editorials selected here, he writes about the need for constant skepticism of the received scientific wisdom, giving the example of a new class of compounds involving the noble gases that were once thought to be unreactive. He ponders the roots of scientific integrity and the dangers of overspecialization, pointing out some of the characteristic strengths and weaknesses of scientists as a group.

One of Abelson's major themes has been the dangers of unrealistic expectations. Thus he is concerned to dampen expectations of unlimited growth in the budgets for scientific research. He warns against overconfidence in our technologically based prosperity, correctly foreseeing some of the strains of the 1970s and 1980s. He calls attention to the myth that medical science will be able to cure anything and points to the importance of a patient's own habits of diet, exercise, and abstinence from smoking.

He returns repeatedly to questions of science education, pointing out the essential connection between a research environment in the universities and effective training of the next generation of scientists. Well before the current concern about science education, he argues for more exposure to science in the grade-school curriculum. He points out that nearly all industrialized countries provide more intensive training in math and science, and graduate more engineers in relation to their population size, than does the United States. He argues for more support and fewer federal restrictions in supporting university education, pointing to the connection between an adequate supply of scientists and engineers and economic competitiveness as a nation.

Early in his editorship, Abelson called for tougher standards in the publication of scientific papers. He championed the right of the public press to print news of scientific discoveries prior to publication in scientific journals, refusing to follow policies advocated by leading physics and medical journals, while calling on the scientific community to discipline scientists who exploited the gullibility of the media. And he called attention to the poor quality of much science reporting in the media, and, in the 1970s, to the flood of pseudoscience books and their distressing popularity on college campuses.

Abelson has been concerned also about the institutions that connect this country to others, and he wrote of the need to maintain scientific interaction with the Soviet Union while taking steps to insure that exchanges were not one-sided. He pointed out a lack of science and technology training in our foreign aid efforts, which he saw as crucial to the self-sufficiency of developing countries.

Finally, he looked to the health of disciplines and of science-based industries. He chided the physics community for what he saw as unrealistic arguments in support of high-energy physics. He worried about the motivation of doctors in the midst of the rapidly changing technological, legal, and regulatory climate of medicine. And he celebrated the rise of small, innovative biotechnology companies as heralding rapid application of new biological knowledge.

19 March 1971

Each week *Science* prints about the same amount of material as do leading newsweeklies. We strive for and usually maintain standards of accuracy that compare favorably with the best of them. Yet we must achieve this with a fraction of the staff engaged by other weekly magazines. We could not accomplish our tasks and maintain standards were it not for the generous cooperation of many scientists. They provide letters, articles, book reviews, scientific reports, technical comments, and meetings reports. More than six thousand reviewers help us in the selection of these materials. This cooperation arises naturally from members of a community who are accustomed to working unselfishly toward constructive ends. Final decisions must be made in our editorial offices, but, in terms of creating material for *Science* and judging it, the community-at-large is a considerably greater factor than is the staff.

Because *Science* is published by the American Association for the Advancement of Science, members rightly consider that the content of the magazine is a matter about which they are entitled to form and to advocate opinions. In consequence, there is rarely a time when the editors are not under various kinds of pressure to publish or not to publish this or that.

The members of AAAS include a heterogeneous group of scientists, engineers, physicians, and others. They represent many disciplines and specialties. They are drawn from academia, government, industry, and private practice. Broad spectrums of ages and political attitudes are also represented. When permutations and combinations of background and interest are taken into account, few readers can be expected to have identical needs or views as to what should be published in *Science*.

One issue is the balance between material that is strictly scientific and material that is concerned with interactions of science and technology with society. Surveys indicate that the majority of our readers are working scientists, often with teaching responsibilities, who are more interested in new developments in science and in review articles than in public affairs. These scientists are often critical of our policy-oriented material. Those who have administrative responsibilities follow such material closely and value it, while paying less attention to the more technical content.

Tensions and emotions are aroused over scientific material, but they are minor in comparison with those arising when political issues are involved. Policy matters can be discussed with good humor and objectivity—as long as they are not in the middle of an active political arena. So strong are the emotions of the political process that the question can be raised as to whether a scientific organization can survive if more than a modest fraction of its activities are oriented to the political issues of the moment.

But the future of civilization is dependent on the wise use of science and technology, and members of AAAS cannot responsibly avoid participation in some of the opinion-making and decision-making processes. *Science*'s policy in this matter is to attempt to give fair representation to the broad spectrum of views. On many occasions the board of directors of AAAS have considered the role of *Science*, and they have reiterated that it should function as a forum rather than as an official spokesman.

Audience participation in the editing of *Science* produces an unending series of problems major and minor. Without that participation, however, life would be dull and the magazine would be neither interesting nor consequential.

12 October 1962

The recent synthesis of xenon tetrafluoride and related compounds makes necessary the revision of many chemistry textbooks. For about fifty years, students taking elementary courses in the subject have been taught that the noble gases are nonreactive. Millions of pupils have absorbed this dogma and faithfully parroted it back at examination time.

The first evidence that xenon might participate in chemical combination was obtained by Neil Bartlett, who suggested that compounds of the type $XePtF_6$ could be made. This discovery has been followed up by a team of scientists at Argonne National Laboratory. The work they present is clear-cut and convincing. Xenon reacts with fluorine to form more than one relatively stable compound. A variety of different procedures independently confirm the chemical constitution of the new product. Indeed, the ease with which XeF_4 is made and its properties are explored is almost shocking. One can introduce the two gases into a simple system, heat the mixture for 1 hour at 400°C, and observe the formation of crystals. The experiment can be performed readily by any chemist and by many other scientists, even though they may have had only elementary training in chemistry. Some caution must be employed, for fluorine is poisonous and reactive, and the xenon fluorides may be dangerous. However, xenon and fluorine are available commercially in safe containers. Thus the essential ingredient in discovering XeF_4 was not money or equipment, but an idea. Even the choice of fluorine as a reactant seems obvious since it is the most reactive of all the elements.

There is a sobering lesson here, as well as an exciting prospect. For perhaps fifteen years, at least a million scientists all over the world have been blind to a potential opportunity to make this important discovery. All that was required to overthrow a respectable and entrenched dogma was a few hours of effort and a germ of skepticism. Our intuition tells us that this is just one of countless opportunities in all areas of inquiry. The

The Need for Skepticism

imaginative and original mind need not be overawed by the imposing body of present knowledge or by the complex and costly paraphernalia that today surround much of scientific activity. The great shortage in science now is not opportunity, manpower, money, or laboratory space. What is really needed is more of that healthy skepticism which generates the key idea—the liberating concept.

---

24 April 1964

One of the most astonishing characteristics of scientists is that some of them are plain old-fashioned bigots. Their zeal has a fanatical, egocentric quality characterized by disdain and intolerance for anyone or any value not associated with a special area of intellectual activity.

This attitude may have its beginnings in undergraduate studies; it is strongly nourished in graduate work. During this period the student is subjected to enormous pressures toward specialization. His or her course work is directed toward a limited area of science. Thesis research is even more strongly focused on a tiny area of inquiry. To achieve a Ph.D. degree a candidate must work hard and spend especially long and devoted hours. Graduate studies demand an overriding priority above any other physical or intellectual pursuit. A graduate student is driven by the situation but he or she also must become a most hard-eyed taskmaster. Every kind of psychological weapon must be used to achieve the necessary concentration of effort. One of the most useful processes is to become convinced that the area of knowledge under study is indeed the most important possible. As a corollary all other intellectual pursuits can be ignored as worthless. It is necessary for virtually all scientists to adopt such rationalizations from time to time. To achieve success one must concentrate on performing a series of specific tasks with complete rigor. Putting the blinkers on is a great help toward this accomplishment. The trick is to know how and when to take them off. One must be able to specialize but one must be able to escape the web of one's own rationalizations. Many have not the will or wit to do this. Thus they are cut off from the rest of the evolving fund of knowledge. For a time such specialization has survival value after graduate school. It can lead to early establishment of a scientific reputation. In the end, however, it is often bitterly self-defeating.

A frequent consequence of bigoted overspecialization is early obsolescence. Areas of science that are at the center of the stage at one time are destined to be mined out in a few years. As the mining process nears completion many concern themselves with ever more specialized and trivial aspects. Ultimately they discover that the rest of the world has passed them by, that few others are even slightly interested in what they are doing. They face the need, first of overthrowing deep-seated prejudices and then of acquiring a whole new body of knowledge and techniques. Few succeed. Some turn sour and in effect die intellectually thirty years before they are buried.

Avoidance of bigotry carries with it important bonuses. If one is tolerant and willing to admit quality in others, the world can be a great teacher. In universities professors give guidance as to what is important and worthwhile. After university days, the scholar has a more difficult problem. He or she must become aware of the existence of an important body of information, select that limited portion for which there is time to absorb, and then study it. Thus one must be both professor and scholar. But if a scholar is tolerant enough the world will become his or her professor. With some effort a scholar can identify a host of others who have wisdom and taste. With their help he or she can enjoy a continuous process of self-renewal.

11 October 1968

Many people view growth as akin to progress, achievement, and the good things of life. In contrast, a steady-state situation is viewed darkly. A 5 percent annual increase in the gross national product is considered healthy, while failure of the economy to attain an increase would be considered a cause for great alarm. The most valued form of growth is a steady increase each year. This can be expressed by the equation $x = x_0e^{kt}$, where x is the variable, x_0 is its value at time $t = 0$, and k is the growth rate. When $kt = 0.693$, $x = 2x_0$. Thus a growth rate of about 3 percent a year leads to a doubling time of twenty-three years. Such a rate seems sedate enough, but as time passes further doublings occur so that ultimately the value of x goes to infinity. In any practical situation this is impossible and, as Platt has pointed out, continuous growth often leads to great problems for society.

In the early part of this century our population growth was a source of great pride, while the more nearly static populations of some European countries were considered indications of decadence. Lately our attitude about population has been changing.

Despite our new realization that some kinds of growth are not good, this lesson will not be applied generally for a long time because of our inherent prejudice in favor of growth. Today the public is becoming concerned about the way nature is being despoiled. However, few seem to realize that most ecological problems can be traced to some aspect of exponential growth. In attempting to prevent further deterioration of the environment, ecologists and conservationists may find that their strategy of piecemeal attack on specific situations wins battles but loses the war. The toughest enemy is the inexorable exponent.

An example of the kind of problem ecologists face comes from the electric power industry. To satisfy public demands, the industry has increased its installed capacity at the rate of 6 to 7 percent per year for many years. Typical projections assume a similar rate of increase far into the future. All of us are indebted to this industry and the conveniences that it brings us. Take away dependable electric power and there remains a shambles. Yet the projected expansion will create great tensions. Already there are siting problems and complaints of thermal pollution. Air pollution and dangers connected with the nuclear industry will increase. At some point society must conclude that an exponential expansion in power output is not desirable.

Scientists are in the midst of traumatic sequelae to an unsustainable exponential growth in the support of research. Over a period of about two decades, beginning in 1940, federal expenditures for research and development rose by about 25 percent per year. When such growth was sustained for some years, the beneficiaries expected it to continue indefinitely. They were inclined to accept exponential growth as a law of politics or nature. Even as recently as a few years ago it was widely held that federal support for science should increase at the rate of 15 percent a year. This was at a time when the GNP was growing at the rate of about 5 percent. Scientists might hope for, and argue for, a rate of growth somewhat larger than the GNP, but the larger the disparity, the quicker the disappointment.

Society has been, and still is, on a great growth kick. If we are interested in a long-term future for man, we will regard rapid growth with suspicion. We will look for, and point out, the unexpected and unpleasant consequences of exuberance long continued, and seek to moderate it before irreparable damage has been done.

21 March 1969

Great achievements often carry with them the seeds of future failures. Repeated success breeds overconfidence and unwillingness to persist in the hard measures that led to excellence. Prolonged enjoyment of excellence brings indifference and even contempt for it. Examples of these tendencies of human nature can be seen in current attitudes toward science and technology.

When people witness accomplishments such as those of *Apollo 8* and *Apollo 9*, they are impressed with the power of American technology. They are inclined to say, "If we can do that, we can do anything." They are also inclined to believe that we can do everything—that, given the goal and the money, technology can be bent to the accomplishment of any and all tasks. This is not true. Technology cannot rescue society from unlimited folly—a long-continued population explosion, for example.

Overconfidence in our technology leads to other faulty judgments. As Lee Dubridge has recently pointed out, we have become so accustomed to the almost magical capabilities of technology that we expect instantaneous solutions to all problems, no matter how complicated. This demand is unreasonable, even when the problems are purely technical. When complex social, political, and ethical considerations are additional important factors, rosy expectations are just plain foolish.

Confident in the power of American science and technology, the public is indifferent to them and turns its attention elsewhere. In response to the clamors of the moment, many of the brightest young students drift away from the physical sciences, seeking a future role in solving social problems. In some instances their youthful enthusiasm may produce worthwhile change, but many will discover that the problems of society are not easily solved and that the hard-won progress of today often disintegrates tomorrow.

After a period of enjoyed prosperity, affluence seems to be a guaranteed feature of life. Why struggle for it? Critics see that

Overconfidence in American Technology

affluence has a seamy side—pollution and the like. And so they criticize, and rightly so. However, being human, some do not stop with constructive criticism. They go far beyond that, asserting that technology is the source of most of our present social problems. Perhaps it is, but who wants the standard of living and the pestilence of the Middle Ages?

We must learn to live more wisely with technology, but we cannot abandon it. We cannot even assume that present technology will guarantee future prosperity. During the past few decades the cream has been skimmed off many natural resources. During the next decades raw materials will become more costly, and scarcities of many items will be an unpleasant fact of life.

At the beginning of this century this country was the world's leading miner of gold; today we produce a small fraction of the total. At that time we had vast supplies of copper ore, with a copper content of about 2 percent; the average copper content of the ore that is being mined today is about a third that much. Once we were a great exporter of iron, lead, silver, and petroleum. Today we are a net importer of these items.

To maintain solvency, we must find new sources of raw materials or develop substitutes, or improve our competitive position in world trade so that imports of raw materials can be paid for. All these possibilities involve—among other things—the requirement that our science and technology be excellent. Prerequisite to the maintenance of such competence are more realistic attitudes toward science and technology on the part of all of us, including the public and their political representatives.

4 June 1971

To serve the public most effectively, scientists, engineers, and physicians need a friendly, sympathetic environment and one in which their limitations are understood. Until recently such an environment prevailed, but the last few years have brought changes in the public's attitudes. Many adverse changes are involved, but one of the most important is the public's inaccurate estimate of what can be achieved. The public has come to expect miracles from engineering and medicine.

Today medical scientists and physicians are particularly subject to pressures produced by unrealistic demands from the public. Failure to produce miracles is taken as proof of a refusal to be interested in meeting the public's desires. One indignant correspondent wrote me in part as follows:

> It would certainly seem from the record to date, that basic research scientists are so preoccupied with the exquisiteness of their protocols and the refinements of their research approaches that they have missed entirely the objective and intent of we taxpayers who are funding their 'ivory tower isolationism.' . . . we as taxpayers expect something more in return than scientific dialogue between the scientists at seminars and in medical journals.

During recent discussion concerning the new cancer program, similar remarks were made. If the critics are unwilling to grant humane motives to medical scientists, the critics should at least recognize the power of self-interest. Anyone responsible for a dramatic advance in the fight against cancer will receive recognition and undying fame.

In some aspects of medicine, physicians can perform today what would have been impossible forty years ago. Yet there are other aspects in which the practice of medicine has not changed much. For example, little can be done to halt the aging process. Neurological disorders are another area in which the medical profession cannot deliver as much as people wish.

Physicians find that many patients have unrealistic ideas of what doctors can do for them. Many people seem to think that they can carelessly expend physical resources in any way they wish and that the doctor can always cure them. Patients give no cooperation in the matter of exercise, diet, or abstinence from heavy smoking and yet expect the doctor to make up for their neglect. They want a drug to substitute for self-control and their own efforts. That kind of miracle is not yet available, nor is it likely to be.

Expectations for magical cures are partly responsible for the growing frequency of medical malpractice suits. If a cure is not forthcoming, the patient assumes the doctor must be at fault. Sometimes a doctor is at fault, but often too much is expected of him or her. So common have malpractice suits become that fear of them is adversely affecting the practice of medicine and is raising the cost of medical care. Today many doctors find it necessary to practice medicine defensively—that is, instead of concentrating only on the well-being of the patient, they must also order a large number of unnecessary tests and examinations to answer questions they may be asked in court.

Scientists, engineers, and physicians cannot realistically hope that the favorable climate of ten years ago will soon be restored. They can foster an improved climate, however, by persistently reminding the mass media and the public of the many limitations of science, technology, and medicine. We are neither witch doctors nor magicians.

18 August 1967

As a nation we often behave as if we are not sure that we will survive the next six months. Urgent short-term or emotion-laden issues commanding the headlines compete successfully for federal funds, while programs essential to the long-term life of the nation are neglected. In such circumstances, academic scientists must not fail to remind the public of the many enduring benefits to be derived from support of research.

The public is aware that practical applications have arisen from past research and are likely to arise from future research, but scientists would do well to continue to furnish examples of the relation of research and beneficial applications. Another need is to help the public explore the cultural aspects of scientific knowledge. Most humans hunger to understand the universe about them, and many are willing to make considerable efforts to satisfy their curiosity. Thus the museums of the Smithsonian Institution in Washington, D.C., draw large crowds. The observatory at Mount Palomar is besieged with visitors, and the Christmas science lectures sponsored by the American Association for the Advancement of Science are well attended.

Leadership in the creation of knowledge brings great national prestige. When a conspicuous contribution is recognized with a Nobel Prize, a nation's stature increases. The United States has been receiving about half of the Nobel Prizes, and most of the winners do their research at universities.

A serious failure of academic scientists has been in educating the public with regard to the role of scholarly inquiry in the universities. The necessity to do so became acute a few years ago. At that time a number of articles in major publications asserted that research efforts by professors were destructive to the teaching functions of universities. Critics neglected to mention that often the most incompetent professors in science departments are those who do no research. The administrations of many colleges and universities quietly responded to the criticisms by making clear to their faculties the importance attached to the teaching function. However, the public is largely unaware of these steps, and an impression remains that good teaching and research are incompatible. This is an incorrect view.

With science evolving rapidly, a major task for professors is to keep up with developments in their field. The full-time instructor who presents material that is out of date defrauds his or her students by failing to render proper guidance with respect to subject material, by failing to set high standards of scholarship, and by failing to inspire enthusiasm for learning. To be a good teacher of science, a professor must be intellectually vigorous. He or she must be part of the creative enterprise. The most practical means of keeping current with new developments is to participate personally in research activity. The sharply disciplining nature of cold-eyed peer evaluation induces research scientists to work hard at creative endeavor. As part of that effort they try to achieve awareness and understanding of new discoveries in their branch of science. Their students are beneficiaries.

A final role of academic research is in the graduate training of scientists for industry, government, and academia. If good basic research is not conducted in the universities, how will the nation obtain the elite scientists so essential to modern civilization?

28 December 1962

Take away science and technology from our civilization and there would remain only chaos and starvation. We exist in complete dependence on an organizational and production complex that provides food, clothing, shelter, and the common defense. Less obvious, but equally important, is the philosophic significance of the knowledge that science has generated. Attaining an understanding of the natural laws that govern our lives and the universe about us is a profoundly enriching experience. Unfortunately, only a relatively few citizens, mostly scientists, understand the implications of science or can visualize its future impact. Some humanists, having only the haziest concept of science, have come to regard it as a mysterious and intractable Frankenstein. Others are more constructive and have discussed the need for communication between scientists and nonscientists and especially between scientists and politicians. The gap between the scientists and other citizens is growing, and scientists will have to assume a substantial share of leadership in meeting the problem. Hence it is timely to present one aspect of the matter. James H. Mathewson has addressed himself to the question of college curricula for the scientist and nonscientist, and he argues thoughtfully concerning the inadequacies of present approaches. He points out:

> . . . elementary science courses are not taught with a broadening function in mind. They are designed to train the science major in specialized fact, theory, and technique from the start. They generally cover only one field in science, with little instruction in how the subject relates to other fields inside or outside of science. Under these circumstances the nonscience major finds his encounter with science a torment of meaningless detail, providing little that he may profitably use for a wider purpose than satisfying an academic regulation. He does not need to become a specialist in a science; he does need to understand the essential nature of science as a whole and his relation to it.
>
> The science major remains correspondingly undereducated. He is frequently permitted to avoid all but the briefest exposure to nonscience courses and activities.

Science and the Humanities

Mathewson proposes revisions of the content of survey courses. We believe that implementation of his ideas would have constructive consequences. But we doubt that his suggestions are sufficiently comprehensive to meet the challenges of the need. First, a quibble about his proposal that the humanists study scientists rather than science. An implication is that there is such a thing as a type specimen, a standard sample, a guaranteed genetically pure "longhair." Actually, in behavior and thought pattern no two scientists are alike. Many, however, are characterized by a hunger for knowledge that does not stop at the boundaries of their specialties. Once their formal education is finished they inquire into other fields. After the rigors of training in science, the subject content of the humanities seems hardly more difficult than a good novel. While it is feasible for a scientist to overcome deficiencies in earlier training it is almost impossible for humanists to acquire a knowledge of science once the formal educational process is completed. An average man or women, even a superior one, cannot learn science from scratch. Our principal comment, then, is that a drastic revision of the educational process, including secondary school training, is overdue. We believe that a realistic curriculum for the secondary schools might well include almost continuous exposure to science, beginning in the primary grades. This would give partial recognition to the realities of a changing world and enrich immeasurably through philosophic values the lives of all.

---

28 November 1980

In comparison with other advanced countries, the United States is becoming a nation of scientific illiterates. Our principal commercial and military rivals have recognized that future superiority will rest heavily on competence in applied science and engineering, and they are preparing their young people for the world of the future. For example, in Japan instruction in science begins in the first grade. From the third through the ninth grades, science and mathematics are required subjects and constitute two of the four major courses taught. Most students who intend to go on to universities continue to take science and mathematics courses in upper secondary schools. Their curriculum includes differential and integral calculus and probability and statistics.

Conspicuous examples of American achievements, such as Nobel Prizes or pictures from the *Voyager* spacecraft, serve to blind the public to the fact that a problem exists here. Actually, the Nobel laureates and the engineers responsible for *Voyager* are products of an earlier era. During their formative years (the 1940s and 1950s) a different attitude, one more favorable to science and engineering, prevailed in America. Around 1965 the environment for science and engineering began to deteriorate, and while Nobel Prizes still come, our superiority in technology has about vanished.

A recent report entitled *Science and Engineering, Education for the 1980s and Beyond* that was prepared by the National Science Foundation and the Department of Education, provides some sobering comments. At a time when the world faces an enormous need for engineers, the United States lags behind Japan, West Germany, and the Soviet Union in the number of engineering graduates per capita. The contrast is especially marked with respect to Japan, where engineering enjoys high prestige and the total number of degrees granted to engineers annually has surpassed that in the United States. In Japan 20 percent of all baccalaureate and about 40 percent of all master's degrees are granted to engineers. This compares with about 5 percent for each of these degree levels in the United

States. Moreover, many of the U.S. graduates are foreign nationals.

In Japan an engineering degree is a favorable route to business and social success. The report states that in Japan "only about 50 percent of the engineers produced each year . . . enter the engineering profession. The others become civil servants and managers in industry. Around one-half of the senior civil service hold degrees in engineering or related subjects. . . . In industry, about 50 percent of all directors have engineering qualifications."

The education situation in Germany is similar to that in Japan, with emphasis on science and mathematics in primary and secondary schools. "The overall picture in Germany is one of a very high level of science and mathematics literacy among college graduates as well as a strong science and mathematics understanding among the general population," according to the report.

In the Soviet Union students are exposed to an intense mathematics and science curriculum. Algebra and geometry are taught in the sixth and seventh grades, and additional mathematics, including calculus, is part of the high school curriculum. All youngsters are required to complete five years of physics and four years of chemistry. About five times as many Soviet students as Americans go on to engineering training. The inefficiencies of the Soviet system dissipate much of this advantage, but one can scarcely feel comfortable about the contrast in educational level between the military forces of the U.S.S.R. and the United States.

Our present policy is moving us toward becoming a colonial supplier of raw materials and food to more advanced countries and is placing us in a position of increasing peril. Unfortunately, there is no crisis to alert the public. The one positive factor operating at this time is a strong demand for engineering graduates, which is driving up salaries. Overcoming scientific illiteracy will take decades.

---

9 January 1981

For the next decade this nation will have great needs for scientists and engineers. Already there are shortages in such fields as petroleum and chemical engineering and computer science. The shortages are likely to spread to other fields and to worsen. Faculty members are being recruited by industry, and fewer students are seeking the Ph.D. degree in some disciplines.

There is a particularly acute shortage in computer science. Advances in electronics have made computation and memory comparatively cheap. Many companies perceive attractive opportunities to apply the new hardware. These applications require development of systems of software by computer scientists. Eager bidding for computer scientists has pushed salaries to $30,000 per year for students at the baccalaureate level.

We are also experiencing an expanding need for scientists and engineers brought on by the decreasing availability of oil and by its higher price. Merely to maintain our limited production rate of petroleum will require employment of more experts for planning exploitation of the new wells and for tertiary recovery. Research, development, and engineering entailed in synthetic fuel plants will employ an even larger number of trained people.

As prices for energy escalate, most of the nation's existing processes, equipment, plants, and buildings are becoming obsolete. As prices for oil go even higher, there will be great incentives for research and development, and design and construction of new facilities. In the chemical industry, the need to minimize the hazards of toxic substances is also leading to major efforts involving highly trained personnel, including toxicologists.

The United States has lagged in achieving standards of quality control. The Japanese, who have been exemplary in this regard, employ a much higher ratio of engineers to blue-collar workers than we do. To identify and correct the factors that lead to manufacturing defects often requires unusual skill and ingenuity. Restoration of the reputation of American products will demand the deployment of more scientists and engineers in production facilities.

There will be other major demands for trained personnel. This country has lost its lead in many areas of technology. Who is to innovate? Who will create and develop alternative energy sources? The world will face terrible shortages of food, and we will need to change our style of agriculture to slow soil erosion. Who will develop more productive and pest-resistant plants? Who will exploit the opportunities of the revolution in biology based on recombinant DNA?

The universities, traditionally a source of new knowledge and trained people, will try very hard to fulfill their function. However, they are in relatively poor shape to do so. In most schools, equipment for training and research is antiquated or absent. Opportunities for young faculty members have become limited. At the public universities, state funds for support of research and teaching have diminished. Flexibility in the use of funds has been curtailed. The gross total of federal research funds in constant dollars has remained about the same. But federal intervention in universities has entailed substantial costs. For this and other reasons, the net sums actually available for research in universities have declined substantially.

One of the few bright spots in the situation is an improving relationship between industry and the universities. Leaders of industrial research are enthusiastic about the quality of the young people they are hiring. An increasing number of companies are supporting fellowships and research at the schools. They could and should do more. They could serve their long-term interests by helping to improve the level of equipment for teaching and research. They should be more emphatic in expressing their admiration for the training that young people are receiving. They should be prepared to intervene if federal budget cutters should propose to deal another blow to universities by chopping federal research funds.

10 May 1963

The scientific community has been curiously flabby in reacting to evolutionary trends that challenge the vitality of science. Modes of communication that were adequate fifty years ago have not been altered, in spite of the vast increase in numbers of scientists. The annual round of spring meetings reminds us that these great national gatherings are losing their effectiveness as media for scientific communication. At the recent Atlantic City meeting of the Federation of American Societies for Experimental Biology there were 3,138 papers presented and as many as thirty-four simultaneous sessions. There are comparable situations in other areas of science. Planning one's program of attendance on such occasions can be frustrating, for one notes numerous papers of interest but discovers that many of the attractive presentations are being given concurrently. All too often the harassed scientist cannot make up his or her mind and forgoes all of the choices.

The proliferation of scientific literature has comparable negative aspects. Faced with a flood of material, no scientist can do more than sample the publications appearing in one immediate field and in the relevant neighboring disciplines. Here, too, each scientist has a breaking point at which he or she gives up on the effort to follow new developments.

Instead of tackling these communication problems we have ignored them, and we have retrogressed, for we have allowed our standards to deteriorate. We permit and even encourage scientists to deliver virtually the same lecture at meeting after meeting. It is annoying and wasteful to make a special effort to hear a paper only to find that the speaker is repeating, almost verbatim, material presented earlier.

This tendency toward repeated presentation has also affected the literature. I have noted instances in which basically the same article has appeared more than five times. This repetition is compounded in the structure of the usual scientific paper. A scientist will obtain one new result, the essence of which can be stated in a paragraph and a table. In the standard min-

Some Needed Reforms

uet, the scientist expands the paragraph to ten pages as he or she describes a new fact in the abstract and presents it again in introduction, discussion, conclusion, and summary. When such a paper is published repeatedly, the author can easily succeed in restating one basic paragraph several dozen times.

The present communication problems could be greatly ameliorated if the scientific community would adopt a tougher standard of what is acceptable. If editorial policies were tightened, the amount of material appearing could be cut to a quarter of the present volume with no essential loss. This tougher approach might well take the form of a stern attitude toward repeated publication of the same material. It would require some reforms in the conventional structure of papers, so that key ideas would not be repeated so many times. It might be necessary to suppress the tendency toward premature publication of fragmentary results.

A parallel toughening in our approach to scientific meetings also would be useful, and the number of sessions could be cut drastically without much loss.

Such needed reforms would have obvious beneficial consequences. To implement them requires courage on the part of editors and officers of societies and generous cooperation and understanding on the part of scientists-at-large.

18 January 1963

The reporting of scientific news has become a controversial matter. Robert C. Toth in the *New York Times* raised a question concerning the release of news of space research—specifically, Mariner II mission results.

> ... Should the discoveries be given to the press as soon as available? Or should they, like other scientific results, be given first to the scientific community as some scientists demand?
>
> The public, whose tax funds financed the experiments, have a stake in the matter.

Toth was unhappy with policies of the American Institute of Physics set forth in an issue of *Physical Review Letters*:

> Scientific discoveries are not the proper subject for newspaper scoops.... In the future, we may reject papers whose main contents have been published previously in the daily press.

The *Applied Physics Letters* of the Institute adopted a similar policy:

> Work described elsewhere, for example, in press releases or in the form of abstracts of contributed papers, prior to scheduled publication in A.P.L., will not be considered eligible for publication.

Recently I have been urged by the American Institute of Physics to adhere to their position. Although generally sympathetic to their stand, I cannot completely agree. The policy seems rigid, and the attitude toward abstracts of scientific papers severe. I feel that newspapers and scientific journals are not in serious competition with each other. These media are worlds apart in audience, coverage, and precision of technical detail. It is true that the volume of news of science in daily newspapers is increasing. In Washington and New York, coverage is excellent: the writers are exceptionally competent, and sometimes adequate space is devoted to their stories. In other parts of the country, science reporting ranges from fair to downright mediocre, or there is none at all. Some good, authoritative material is provided by the wire services, but local editors butcher it with a heavy hand. The material that is printed is usually gee-whiz, Buck Rogers distortions of the facts. Science writers for the wire services, wanting their copy to be used, tend to seek the more glamorous items. With distressing frequency scientist-operators are able to flimflam the science writers with news stories that excite the imagination but have no solid technical basis. Local editors are especially susceptible to these worthless baubles, which they run in preference to less exciting items of solid merit.

The alert scientists gives only limited credence to newspaper stories. He or she finds them valuable as indicators of important events. To obtain full details and sufficient information to judge the validity of a claim, the alert scientist consults the scientific literature.

It is tempting to try to reform operators by denying them publication in scientific journals. This mechanism would create a distortion of the true functions of journals. Other, and better, mechanisms are available. The major responsibility properly rests at the local level—with scientific colleagues, with superiors, and with properly constituted news bureaus.

31 May 1974

A cursory examination of news sources leads to the conclusion that citizens have a great number of opportunities to become well informed. They can view programs on the various television channels. Most radio stations give the news at least hourly—some continuously—and there are many talk shows. Newspapers and newsweeklies attempt to carry on their traditional function.

Yet a closer examination reveals that the news media are not effective in presenting balanced news in-depth, but are to a degree contributing to a malfunctioning of society. They have participated in creating and exacerbating a series of crises by overconcentrating attention on particular topics. Typically, after a period of concentrated attention, the media suddenly drop one topic as they rush to indulge in overkill of the next one.

These tendencies were noted by Alan L. Otten in a recent column in the *Wall Street Journal* that began:

> One hallmark of contemporary America, it's frequently been noted, is the short life-span of its crises.
>
> A problem emerges suddenly, builds swiftly to crisis proportions, briefly dominates public consciousness and concern, and then abruptly fades from view. Civil rights, urban decay, hunger, drugs, crime, campus unrest, medical care, the environment, energy—one succeeds another with blurring speed, almost as though some issue-of-the-year club were in charge.

A glance at Otten's list leaves one with the impression of a variable amount of residue from the periods of great mass media attention. Most of the topics listed are now practically dead as far as the media are concerned. True, there is a considerable residue from emphasis on the environment both in legislation and in public consciousness, although with sharply curtailed media coverage, the public concern and interest have lessened. After tremendous attention, news coverage of the energy crisis has almost disappeared, and there is little indication of substan-

Media Coverage of Substantive Issues

tive progress in meeting the issue. The basic problems remain, but the public is bored with the subject, and the net effect of the coverage is to make it more difficult for progress to be made in the future.

Another undesirable feature of the massive attention is its lack of quality. The bizarre and the spectacular news takes precedence over reports with balance and substance. We at *Science* frequently have opportunities to evaluate the performance of the media in unearthing the facts about a given situation, and more often than not we are disappointed. This is particularly true in those areas in which science and technology interact with public policy. These issues are usually complex and enduring and not well handled by slick or sensational journalism.

The current practices of the mass media point up the value of publications like *Science* that are designed to inform rather than to excite. Although our resources are comparatively modest, we feel no handicap in competing. On any topic we choose to cover we can, if we wish, produce a more rounded, complete, balanced, and scholarly story. Usually we do not choose to compete on topics that are being well covered by others. We prefer to pinpoint issues before they are in vogue, and we are not averse to dealing with significant topics after others have dropped them, provided there is new and relevant information.

In our efforts to maintain quality, we are fortunate in having a readership that expects good performance. Our authors understand this and tend to behave accordingly. We are also fortunate in having an audience that values rigor and discussion in depth and is willing to contribute ideas, time, and money to the common objective.

21 June 1974

During the last few years elements of the public, and particularly of university students, have turned increasingly to mysticism and to what I would call pseudoscience. The top sellers at campus bookstores have included such books as *Chariots of the Gods?*, *Gods from Outer Space*, *Limbo of the Lost*, *The Secret Life of Plants*, and others like them.

The recent pseudoscience books are in part a form of science fiction, but they have characteristics that make them different. The readers of earlier works generally understood that they were scanning fictional material, but the new books seek to create the impression of scholarship and verity. *Chariots of the Gods?* does this in several ways. It has a bibliography. In an introduction it acknowledges help from personnel of the National Aeronautics and Space Administration, including Werner von Braun. The book also contains some respectable scientific material. But the author moves quickly and without warning from fairly solid facts to unsubstantiated speculations.

Another tendency of the pseudoscience books is to evangelize in behalf of fantasies and in the process to denigrate science. For example, in the best-selling *The Secret Life of Plants*, the authors state "what makes plants live, or why, does not appear to be the purview of science." They describe botany as being "reduced to a dull taxonomy."

This is, of course, untrue. One of the great scientific frontiers today is research in plant biology.

The scientists of the so-called establishment are berated because they did not accept the suggestion that plants were capable of emotions that "might originate in a supramaterial world of cosmic beings which, as fairies, elves, gnomes, sylphs, and a host of other creatures, were a matter of direct vision and experience to clairvoyants among the Celts and other sensitives."

In *Limbo of the Lost* the author devotes most of the book to an enumeration of disappearances of ships and planes in the general area of the Bermuda Triangle. In a concluding statement, the author gives his explanation for the information he has produced. He ties the disappearances to unidentified flying objects and concludes that a large ocean vessel and commercial airliner were "actually being taken away from our planet for a variety of reasons."

Much of the appeal of the new pseudoscience seems to relate to a deep-seated quirk of human nature—a predisposition to believe in the supernatural. Part of the appeal of these books is that they are entertaining, interesting, and even exciting. The danger from them is that uncritical and undiscriminating minds may accept imaginative speculation as fact. An optimist might take the view that the current craze for this new form of science fiction will go away, just as streaking departed. But already these types of books have been in demand for several years.

The popularity of pseudoscience books at universities should be a source of concern to academic people, particularly scientists. The new trend comes at a time when many universities have abandoned requirements that students be exposed to as little as one science course. It is not pleasant to contemplate a situation in which our future leaders are being steeped in fantasy and are exposed to a put-down of science without effective response. The university community has a special obligation that it has not been meeting very well. It should move toward providing antidotes to the new intellectual poisons. In meeting these challenges to rationality, we should all remember that although humanity is eager to accept mysticism, it is also capable of yearning for truth.

29 September 1978

The issue of human rights has lent urgency to an assessment of scientific relationships between the United States and the Soviet Union. Such an examination was already in progress before the events of this year and the Kaysen report for the National Research Council was a major contribution. The study focused on a program of scientific exchanges sponsored by the National Academy of Sciences (NAS) together with the Soviet Academy of Sciences since 1959. The program had been created with four goals in view: to establish individual and institutional contact with the scientific community of the Soviet Union; to learn about Soviet strengths and goals in science and engineering; to contribute to improved U.S.–Soviet relations; and to achieve at a later date the "normalization" of scientific contacts between the two countries. The panel noted that the NAS could claim "striking success" in meeting the first three goals.

In a world where thermonuclear war is an ever-present menace, progress toward the second and third goals is of great importance. Nevertheless, in our relationships with the scientists of the Soviet Union, we should not allow one-sided arrangements to persist.

For two decades, the Soviet Union has enjoyed a status of equality to the United States in international scientific matters. The reality is that in only a few fields does equality prevail. The Soviet Union can compete in almost any of a limited number of specific fields that the Kremlin chooses to emphasize. However, the planners are usually years behind the times and world science is conducted on a thousand frontiers. The Russians have never achieved an instrumentation industry and thus most of their scientists are poorly equipped. Another handicap is poor communication. The restricted interaction with the rest of the world has a counterpart in limited interchange within the Soviet Union. Finally, excessive weight is given to party loyalty and to the heads of laboratories, whose tenure is for life. In consequence, young scientists encounter many frustrations. With such

United States–Soviet Scientific Relationships

handicaps the Russians will continue to lag behind in most fields.

Interactions with the Russians occur in a number of ways, such as the Pugwash conferences, academy exchanges, and activities sponsored by the international scientific unions and by UNESCO. The meetings of the unions have involved the most individuals and the greatest interchange of scientific information. Many warm friendships have begun at such gatherings.

The organization and conduct of a large international meeting is a huge task. Almost invariably the organizers find that by far their worst headaches come from the Russians. Many of them send in abstracts and announce their intention to participate. But when the time comes perhaps half will be permitted to attend, thus leaving gaping holes in the schedule of papers. In other instances a group of uninvited or unscheduled people will show up demanding space on the program. The paper of an invited distinguished scientist will often be read by a party hack. When the international meeting is held in Russia there are usually visa problems. The international scientific community should not tolerate such forms of behavior.

In the bilateral exchanges the record is better, and in some areas of science there is a clear gain for both parties. For example, cooperation in the earth sciences has been worthwhile. But too often in other areas the United States has given far more than it has obtained. Current opinion in Washington is that we should be more selective in our interactions. We should ask with respect to a particular field, How good are they? Where are the key installations? Will we have access to the top people and laboratories?

It is in our national interest to continue to have interactions with the scientists of the Soviet Union. But the time has come to conduct the relationship on a tough-minded basis. In the process, though, we should remember that some sanctions may injure well-meaning scientists far more than they irritate the Kremlin.

---

25 March 1966

There is a proverb to the effect that an almsgiver throws a starving man a fish, whereas a truly charitable man gives him a hook and line. The U.S. foreign aid program is in effect almsgiving. We have not recognized what has to be done to bring prosperity to the underdeveloped nations.

In a recent issue of *Science*, Homi Bhabha delineated the problem and suggested a means of solving it.

> What the developed countries have and the underdeveloped lack is modern science and an economy based on modern technology. The problem of developing the underdeveloped countries is therefore the problem of establishing modern science in them and transforming their economy to one based on modern science and technology.

Bhabha went on to advance the thesis that the problem of establishing science as a live and vital force in a society is an inseparable part of the problem. Bhabha spoke authoritatively, for he was the key man in creating a self-reliant atomic energy industry in India. In 1943 no scientific institution in India had facilities for work in subjects at the frontiers of physics, including nuclear physics. Bhabha persuaded an industrialist, J.R.D. Tata, to establish the Tata Institute of Fundamental Research, which has been a constituent institution of Bombay University from the beginning "and has had close relations with many other universities in India, so that students of many of them have done work for the Ph.D. at the Institute." From a small start with a recurring annual budget of $16,000, the Institute grew initially at the rate of 30 percent per year. Its current budget is $3 million.

Early in its history the Institute had a key role in the development of atomic energy.

> . . . groups were established at the Institute to design and build all the electronics instrumentation without which atomic energy work is impossible. Thus, the Physics Division and the Electronics Division of the Atomic Energy

> Establishment at Trombay were both initially housed and built up in the Institute. The electronics group of the Atomic Energy Establishment has today a staff of over thirteen hundred people and is the strongest research and development group in electronics in the whole country.

In contrast to the fine performance in atomic energy, where a strong base in fundamental physics existed, Bhabha described the dismal performance of the steel industry. In turn, German, Russian, and British consultants have been called in, but India still does not have the capacity to design and build new steel plants.

> . . . Unless powerful scientific and engineering groups are established during the construction and operation of existing steel plants as a matter of deliberate policy, the dependency on foreign technical assistance will continue, and the steel industry will not reach a stage of technical self-reliance. A similar situation exists in almost every other industry.

Had Bhabha lived and had his influence expanded, many of India's problems might have been solved. Science and technology can expand faster than populations, thus providing time to solve the population problem. Bhabha and the Tata Institute have shown the profound effect of small sums spent wisely in support of fundamental research. If the United States wishes to be a true friend to the underdeveloped countries, it will find means of helping in the establishing and supporting of indigenous fundamental research institutes. Basic research is only one of several important prerequisites to obtaining optimal benefits from science, but competence in research provides a base from which the most complex technology can evolve when governments are alert and stable.

11 August 1972

During the past several years, the National Academy of Sciences has issued reports on the status and needs of various branches of science. A voluminous report on physics has just been released [A. Bromley, ed., *Physics in Perspective* (National Academy of Sciences, 1972)]. Some of the best resources that the Academy can bring to bear have been mustered in support of these efforts. A substantial fraction of the most competent and distinguished scientists in various fields have labored diligently on committees and panels preparing the materials. Drafts have been reviewed by the prestigious Committee on Science and Public Policy. Typically, in a foreword, the president of the Academy has bestowed his blessing.

The reports have shared many features. They have portrayed opportunities (often interestingly and imaginatively), decried the limited level of support for the particular science, predicted disaster if present trends were continued, and offered the wisdom that most problems would be solved if more money were made available.

Since the reports have been so obviously self-serving, it is not surprising that the prodigious efforts devoted to them have come to little, and it is unlikely that the latest edition will fare much better. This is regrettable, because an excellent case could be made for maintaining the vitality of physics. Within its pages, the report contains interesting and persuasive material. Chapters devoted to the "Nature of Physics" and "Physics in Perspective" are particularly worthwhile. The latter chapter provides an especially good survey of the status of various branches of physics. Some of the potential impact of this material is lost in the excessive length and scope of the remainder of the report. Credibility is also strained by occasional unconvincing arguments and statements. The committee seemed unable to be completely objective in its treatment of high-energy physics. After more than twenty years and the expenditure of more than a billion dollars, high-energy physics has had limited direct impact on other areas of science and, indeed, on the rest of physics. In contrast, low-energy nuclear physics and the physics of energies of 100 volts and less has had very great impact. This impact has taken the form of interaction within physics, usefulness to other disciplines, and many technological applications. Examples of recent contributions of enormous value are laser developments and microelectronics.

Experimental technics and equipment developed by physicists have often been applied in other fields. Indeed, new instrumentation is one of the most important factors in the vitality of the natural sciences. Another major kind of contribution has come from the migration of physicists into such fields as astrophysics, geophysics, and biophysics. A substantial fraction of the nation's best geophysicists received their basic training in physics. High-energy physicists have been among the migrants, but their training has not equipped them so well to be creative in other fields.

Although the committee report fuzzes over the situation, the reality is that there are two quite disparate kinds of physics and only one is highly relevant. Both types should be supported, but in a crunch, high-energy physics should defend itself on its own merits.

The procedure of asking representatives of a discipline to prepare material on their own field has some merit. But the experience of many years and many reports bears out the bromide of not asking the fox to guard the henhouse. Surely the Academy can improve on that procedure.

6 June 1975

A generation ago, the practice of medicine was very often ineffective. Today, because of medical research, diagnosis and treatment are greatly improved. However, the government is now deeply involved in the financial and educational sides of medicine, and a deeper involvement, including detailed management of treatment, is likely. We will probably witness episodes in which well-intentioned but shortsighted governmental action leads to long-term destruction. The academic community, which lives with wreckage resulting from the exercise of federal power, should monitor developments in medicine closely. Intervention may become necessary.

The essence of the practice of medicine is in the interaction between patient and physician. In spite of the tools that have been invented or will be devised, medicine will remain an inexact science. The best physicians are highly motivated, highly intuitive. There is no substitute for the conscience of the physician. If conscience and motivation are lost, little will be left. Current developments are placing at hazard these key factors.

Intervention by the Congress and by the administration has come because of demands of the public that are based, at least in part, on unrealistic expectations of what can be delivered in the way of patient care. The average person's concept of what is possible medically is conditioned by a memory of miracle drugs and polio vaccine and by accounts of organ transplants and great new medical discoveries. The public expects the best possible medical care but wants it delivered in the style of a generation ago—the doctor appearing at the home with black bag and stethoscope. Some medical problems can still be handled in the home, but to do justice to serious illness, the doctor must be able to employ a full set of modern diagnostic aids and therapeutic equipment. The public also has come to demand that physicians never make mistakes in techniques or judgment, as indicated by the current rash of malpractice suits.

Some of the complaints of the public are legitimate. One is the comparative scarcity of general practitioners or primary care physicians. The proliferation of knowledge arising from research

Changing Climate for Medicine

has made specialization in medicine seem necessary. Specialization has had the further effect of encouraging concentration of doctors in big medical centers. The result is geographical maldistribution, with rural areas and the ghettos suffering shortages of physicians.

Without adequate analysis, someone dreamed up the theory that the cure for such shortages was to increase the output of the medical schools. It was reasoned that some of the excess doctors would spill over into the shortage areas. The medical schools were treated to the carrot-and-stick approach. They were put under pressure to increase enrollments and given the inducement of capitation grants, that is, subsidies based on enrollment. To a degree, the treatment has worked. Enrollment today is about 60 percent above that of seven years ago. However, the graduates have settled in the areas that already enjoyed ample numbers of physicians. Instead of getting at root causes of the problem and offering substantial incentives to practice medicine in less desirable environments, Congress is now considering legislation that would force young doctors to spend two years in what amounts to indentured servitude. Moreover, the medical schools now rightly fear that capitation will either be eliminated, causing severe financial distress, or be used as a weapon against the students. One form this action may take is government-imposed quotas on the number of specialists that may be trained. Is the government so wise and foresighted that it can mandate intelligently how many scientific specialists of any kind there should be?

Medicare and Medicaid have also given the government financial power that seems destined to be used to regulate all phases of medical practice. Such details as the length of stay in hospitals are to be prescribed. There is danger that procedures will be standardized and routinized to such an extent that the quality of the practice of medicine will decline.

In these momentous developments, the most important factor of all is being overlooked. How will they affect the motivation and conscience of the physician?

---

11 February 1983

We are now in a period of especially rapid progress in applied biology. Important useful advances have already occurred employing recombinant DNA and hybridomas. Synthetic human insulin is being sold commercially and other major pharmaceuticals for human or domestic animal care are being tested. Antibodies produced by hybridomas have been approved for diagnostic use. Prospects are excellent that viral diseases soon will be conquered by use of interferon or vaccines.

Key ingredients in the dynamism of applied biology are more than 150 small companies, many of them new. Most of them were formed several years ago at the time of the great excitement over the then untapped potentials of recombinant DNA and hybridomas. Some of the companies have already gone bankrupt and others will disappear. A few months ago most observers guessed that there would be a further great mortality of other companies. But prospects of survival have improved.

Genentech is generally considered to be the leading new company. It is a south San Francisco firm that has pioneered in the creation of about a dozen protein products by recombinant DNA techniques. Employees number about 350, of whom 70 have Ph.D.s. The budget for research and development is $21 million. This is small in comparison with the budgets of larger companies, some of which spend ten or more times as much. Yet, in its creation of new major products, Genentech has a record that no other company in the pharmaceutical business has matched in recent years. In part this success is due to the fact that Genentech was early in applying recombinant DNA to create new products. In part success has arisen from its judicious choice of projects to tackle. But probably most important have been the company's policies with respect to personnel, which enable it to attract and retain high-quality people. The best features of an academic environment, including encouragement of publication, are retained. Scientists have equity positions in the company.

Other smaller companies have also succeeded in establishing their own special enclaves in which loyalty and creativity are fostered. In ordinary circumstances at universities, in government, or in industry, a scientist typically manifests only a small fraction of his or her potential. This is due to distractions, multiple responsibilities, interruptions, personality clashes, conflicts with management, and less than complete motivation. An organization that can foster a culture that brings out the best in its people can outdistance its rivals. A number of the new companies are succeeding in doing so. Their rate of progress is now comparable to that of Genentech.

Synthesizing a new product on a laboratory scale is only a short step toward marketing a profitable product. The process must be scaled up, costly clinical tests performed, clearance obtained from the Food and Drug Administration, and then the product must be successfully marketed. These steps require four years or more and ten or more millions of dollars. But there are other ways to obtain a faster financial payoff from new techniques or knowledge. There are diagnostic aids, specialty chemicals, and items for animal care. Many of these items are small in volume but high priced. There are fees for contract research and potential royalties from patents. The successful small companies are carefully selecting viable and limited ecological niches in which they can survive and grow.

The big pharmaceutical, chemical, petroleum, and other industrial firms are intrigued by the potential of biotechnology. They believe that their financial strength, production skills, legal capabilities, and marketing know-how will later prove essential. Many of them are slowly building up their internal research competence. But, in the meantime, the small companies will be moving ahead rapidly to exploit the potentialities of the knowledge base and to extend it. They will be important engines of progress, crucial in establishing and maintaining a fast tempo for the biological revolution and its applications.

29 March 1963

Part of the strength of science is that it has tended to attract individuals who love knowledge and the creation of it. Just as important to the integrity of science have been the unwritten rules of the game. These provide recognition and approbation for work which is imaginative and accurate, and apathy or criticism for the trivial and inaccurate.

The scientist can find many satisfactions from a new discovery. First there is growing recognition of a new truth. This is the most exciting and personally rewarding period. In contrast, the necessary confirmatory work is likely to be drudgery. Another reward can be the approbation that may attend revealing the new truth to professional colleagues. Later comes publication, followed by requests for reprints. To receive a note of appreciation from an unknown reader halfway around the world is a warming experience. Ultimately, it is possible to see the truth incorporated in textbooks as a fully recognized part of the intellectual treasure of mankind.

The rewards have added significance insofar as they are in contrast to the punishments for failure. If success in research comes after a period of barrenness, the accomplishment seems even more exciting. If one has given a talk that has drawn half-hearted response or overt criticism, he or she values good response more highly. After a manuscript has received a scorching review, smooth acceptance on another occasion seems worth a celebration. Those who have published work rightfully castigated for inaccuracies not only experience acute discomfort but serve as a warning example to others.

The quiet personal satisfactions of work in the laboratory are important to the individual. Research, however, is just a pleasant hobby unless its results are evaluated and incorporated into the total body of knowledge. Thus it is the communication process that is at the core of the vitality and integrity of science.

Scientific meetings are often thought of as means of learning of new developments. There is another aspect fully as im-

The Roots of Scientific Integrity

portant that usually is overlooked. That is the effect of a verbal presentation on the speaker. If the event is definitely scheduled some time in advance, the impending occasion can act as a tremendous stimulus. It can cause the investigator to focus more sharply on a particular area. As the time approaches he or she tends to devote all waking hours either to research or to thinking about the topic. The speaker is likely to consider very deeply the limits and certainty of his or her knowledge, to be more self-disciplined, and to do crucial experiments not thought of before or only considered halfheartedly.

A similar series of effects accompanies the writing of a scientific paper. The author quickly discovers how little he or she knows, the gaps that must be filled.

The system of rewards and punishments tends to make honest, vigorous, conscientious, hardworking scholars out of people who have human tendencies of slothfulness and no more rectitude than the law requires.

When the game is played under different rules in an arena such as politics, it should not be surprising that the performance of scientists sometimes leaves something to be desired.

Enough of Pessimism
Designed by Tom Suzuki
Composed by Unicorn Graphics
in Times New Roman
with display lines in Futura
Printed by The Haddon Craftsmen, Inc.
on Warren's Olde Style
Smythe sewn and bound by
The Haddon Craftsmen, Inc.